# GOD, THE BI
# BUNSEN-BURNING ISSUES

**To**

**Ken**

**(Biology teacher and my dad)**

By faith we understand that the universe was formed at God's command, so that what is seen was not made out of what was visible (Heb. 11:3)

# Contents

# Acknowledgements

There is no formula for thanking people who have helped make this book what it is, but I would like to start with the scientists. Exceptionally busy people, often out of the country for weeks at a time, they have been enormously generous with their time, gracious in their sharing of knowledge and swift in their correspondence. Through a series of one-hour personal interviews and email exchanges with real experts, I am better educated. Thank you.

One of the things I have learned from these scientists is that they are firm believers in collaboration. They do not work in isolation – either from the real world or from their professional colleagues. They are team players.

I am delighted to record the fact that in this respect I, too, follow the scientific method.

As Editor of *The War Cry* – the weekly evangelical newspaper of The Salvation Army – I am resourced, encouraged and corrected by a committed and gifted team, who contribute their skills, insights and questions to every article I write, including those articles which make up this book.

My thanks, then, go to: Philip Halcrow, Hayley Pearson, Stephen Pearson, Bruce Tulloch and Christine Clement, for their tireless proof-reading of these interviews; to Gill

Cox, who through her artistic talents made sure these interviews looked great in *The War Cry*; and to my secretary, Joanne Allcock, who transcribed hours of interview recordings, knowing that most of her work would never make the page.

I am grateful to be ceaselessly supported by a further team – the home team. My wife, Maggie, is my greatest supporter, encourager and my inspiration. Our children – Janine and Andrew – are my continual reminder of the wonder, mystery and open-handed generosity of God.

On the subject of genetics, I'm thankful for the unfailing love of my parents, Ken and Jo.

A special word of thanks to the Reverend Professor Richard Burridge, Dean of King's College London, for checking my Bible investigations.

And finally, thanks to my editor Ali Hull, Kath Williams and the team at Authentic Media for their input into this collaboration.

# Foreword

## Rt Hon the Lord Mawhinney, Kt

Professor Richard Dawkins has written 'The existence of God is a scientific hypothesis like any other.' As a trained scientist – BSc (Hons) in physics and a masters degree and doctorate in radiation biology – who taught science to graduate and medical students in the USA and England for more than fifteen years, I do not agree with him.

As a Christian I personally know the God of creation and science through a saving faith in Jesus, who is himself God. No open-minded person doubts the historicity of Jesus of Nazareth – his life, his death, and the transforming power of his followers, acting in his name from within days of his death. Many of his contemporaries testified to having met a resurrected Jesus.

As a parliamentarian of nearly thirty years standing – as an MP, a member of the Cabinet, party chairman and now as a peer – I know that human beings are more than simply an agglomeration of molecules. I have seen demonstrations of human goodness beyond credible explanation other than that men and women are created in the image of God. And I have seen evil which can only be explained by those who believe in human sin.

No, I do not align myself with Professor Dawkins. My experience, my faith and my scientific training and instinct line me up with the scientific contributors to this book. I commend their rational and persuasive contributions. Even if you do not buy their basic Christian presuppositions, they are difficult to dismiss out of hand.

I have never been overly impressed by the arguments of those who try to persuade themselves and others that science and God are competitors or mutually incompatible. Science helps us to understand the 'how' or the 'mechanistic why'. Science does not answer the more fundamental human questions about truth, significance, the meaning and purpose of life or, indeed, death.

These interviews may not give you the answers you may be seeking, but they do point toward a direction of travel for those who genuinely wish to understand themselves in a more complete framework and set of relationships than science can ever contemplate, much less provide.

Scientists and Christians do have at least one thing in common. They both set store by evidence. As one scientist has put it – we are thinking God's thoughts after him. And the range of evidently transformed Christian lives, daily observable, takes some explaining other than by believing in the power of God the Holy Spirit.

*Brian Mawhinney*

# Preface

I am not a scientist. I dabbled a bit as a child. For a primary school project I once made a model of a coal-mine. My dad provided the ingredients for lighting – a battery, a couple of bulbs and a length of wire – and showed me how to connect them so that the bulbs lit up. It was going to stun my classmates and impress my teacher.

The problem was, in class – too keen to show off – I connected just one bulb to the battery. The bulb blew, the whole show failed and I never took to physics again.

At grammar school, despite the chance of eventually getting to dissect dogfish and bulls' eyes, I wrote off biology as a girls' subject – mainly because a lot of girls chose it as their O-level science. (Quite why I didn't consider this an attractive proposition I still haven't worked out.)

Chemistry, on the other hand, was where the action was. Potassium and sodium bursting incandescently when dropped on water. Bangs and flames singeing eyebrows when gases exploded. Mass evacuations when we got too enthusiastic with the rotten-egg hydrogen sulphide experiments. (This was way before today's goggled health and safety regulations, I should point out.)

When it came to A levels, I was seduced by the social sciences of sociology, economics and law. At teacher training college, I studied sociology and – not wanting to write yet more essays – mathematics.

By the time I was teaching maths for a living, I was already regretting not having studied maths and science more deeply when I had the chance.

Two years later, I entered The Salvation Army's training college in answer to God's call to be a minister of the gospel. Since then life has been too full and fulfilling for me to be anything else.

This book is a collection of interviews I carried out for a series in *The War Cry*, the weekly evangelical newspaper of The Salvation Army. *The War Cry* has one main purpose: to present the Christian faith as life-changing and life enhancing. Many of our readers are not churchgoers or yet Christian. Some of our readers are believers and want to be informed about difficult moral and social issues, so that they can better share their faith with others in the technologically aware twenty-first century.

Why am I telling you this? Simply so that you'll know something about where the questions I've asked the scientists come from. They are layman's questions. They come from someone who has picked up his science in the same way as most readers – a bit at school, a bit from books, a bit from the newspapers, a bit through the kids and a bit from the telly.

As a journalist, I think they are questions *War Cry* readers would want to ask if they had the chance.

They are questions that come from someone who believes there are no areas of human experience that God is not interested in – that no subject is taboo to God.

They come from someone who believes that God gave us brains and that being a Christian should not mean being expected to commit intellectual suicide.

They are also questions from somebody who recognises God-given intellect and the personal responsibility each of us has to invest in the talents God has given.

And they are the questions of someone who still wishes he'd worked harder at school.

*NKJB*

# Introduction

There is no escape from science. No matter how inattentive we may have been in school chemistry lessons, we can hardly have failed to notice that the appliances of science have got us surrounded.

Newspapers and TV reports are full of science, from green issues to blue tongue. Stories of a missing child nudge us into forensics. Reports about identity theft or missing government databases introduce the world of cryptography and the magic of prime numbers. Accounts of turkeys with bird flu and cattle with foot-and-mouth are liberally seasoned with servings of veterinary science.

Science is headline news. Barely a day goes by without a science story in the media. A small sample: 'Scientist: Apes are good for a laugh' (*The Sun*: 2 January 2008); 'How going green will add £730 to energy bills' (*Daily Express*: 24 January 2008); 'Scientists create the world's first artificial genome' (*Daily Telegraph*: 25 January 2008); 'MPs back hybrid embryo research' (*BBC News*: 19 May 2008).

Concerns over global warming (geophysics), human cloning (embryology) and the production of GM food (bioscience) are regular ingredients in our media diet. And that's before we start talking about our national obsession – the weather (meteorology). The fact is that if we want to connect with the world around us, even the least promising of school-day pupils has to know a bit of science. Even if we could never tell our carbonates from our carbolics, these days we are having to become aware of our carbon footprint.

Science is served as entertainment. We get regular doses from television medical dramas. Series such as *Casualty*, *ER* and *House MD* open our eyes to a world of clinical technology and terminology that were never found in *Dr Finlay's Casebook* or *Emergency Ward 10* some forty years earlier. Crime dramas, too, are getting back-up from science. In recent years, *Waking the Dead*, *Silent Witness* and *CSI* have raised the profile of forensics and the viewing public, it seems, can't get enough.

Science is also used to sell. From fictional white-coated boffins trying to flog us washing machines to celebrities voicing-over beauty products, the multi-million-pound advertising industry turns to science. The line seems to run something like this: If we can pitch our shampoo/'friendly' bacteria yoghurt/washing powder/dieting product/whatever with scientific-sounding words and images (and therefore 'prove' that it's a great product) then our bottles and packets will defy gravity and fly off the shelves.

Even without drifting into the parallel universe that is the genre known as science fiction, we are becoming more science savvy.

Science is about trying to make sense of the world. Sometimes this occurs on a small scale – so small the human eye cannot see – where scientists describe the activity and possibilities of subatomic particles. Sometimes science is done on a huge scale – so huge the human mind cannot take it in – where scientists talk about us living in an ever-expanding cosmos, where life on Earth depends upon the existence of millions of other galaxies.

Science – the appliance of science – is about solving the problems of the world. Two of the greatest challenges facing the world today are, on a huge scale, global warming and, on a more personal level, the use of genetic technology. Science tells us about the cause-and-effect of burning carbon-based fuels for the past two hundred years. Scientists are developing technologies and promoting strategies to save the planet. In recent years, scientists have decoded the 3.1 billion characters that make up the book of life that is the human genome. Scientists are beginning to prevent disease by developing genetically modified vaccines and to cure conditions through gene therapy.

What are those who do the looking, recording and problem-solving – the scientists – like? According to one stream of popular culture, for years 'mad' and 'scientist' have gone together like 'greenhouse' and 'gas'. For example, H.G. Wells's novel *The Invisible Man* (1897), tells the story of Griffin, a scientist who makes himself invisible but then goes insane when he can't reverse the process. The 1931 film of Mary Shelley's novel *Frankenstein* depicts him as a monster-creating madman. *Dr Cyclops* (1940) features the deranged Dr Thorkel who shrinks four explorers when they threaten to stop his experiments. In the original cartoon *Superman* film (1941), the hero fights a mad scientist who is destroying Metropolis with an energy cannon. World domination is the aim of Ian Fleming's megalomaniac Dr No.

If not mad, then scientists are often depicted as eccentric, as seen in such extreme cases as Sherman Klump in *The Nutty Professor*, Dr Bunsen Honeydew – and his hapless assistant Beaker – in *The Muppets*, time-travel inventor Doc Brown in the *Back to the Future* trilogy and the Time Lord himself, Dr Who.

At the risk of disappointing the reader, it should be pointed out that this book contains no mad scientists or nutty professors. What it does contain, however, is a collection of insights on the complementary roles of science and Christianity in making sense of our world – past, present and future. (Readers will note that all the scientists in this collection are male. When I put this fact to one of their number, his response was that 'top science tends to be male-dominated because it is very competitive, which seems to fit the male psyche.')

These insights are gathered against a backdrop of challenge for people of faith. Just as science increasingly fills news bulletins and entertainment schedules, so science is sometimes promoted as the saviour of humankind's ills. Society looks to science to solve all its problems – from before the cradle (with IVF and embryonic stem cell research) to beyond the grave (through the freezing of human eggs and sperm for use after death).

Science is called upon to create, improve and lengthen life. Science is invoked to protect us from harm, to heal our diseases, to feed the hungry, to free the poor, and to liberate us from responsibility.

With science having all the answers, so the argument goes, then, in a reworking of Friedrich Nietzsche's nineteenth-century maxim: God is dead – and science proves it.

The scientists in this book do not agree with the world-view that God is dead and science reigns supreme. They believe that faith is also a viable method of interpreting the world. They recognise that a faith perspective has been helping people make sense of life and death long before science came on the scene. Collectively, they point out that although science may be omnipresent, it is neither omnipotent nor omniscient. Together, they affirm that God – the Creator, Sustainer and Governor of all things – is.

The scientists in this book face the technological and moral challenges of twenty-first century science. They do not agree with the thesis that if something is possible then it is allowable. They believe that science has moral and spiritual roots. They believe that the appliance of science has moral and spiritual implications.

Do not expect to agree with everything these professors of faith say. As one of their number puts it: 'Scientists see the world slightly differently.' How eccentric, or otherwise, that makes them, I am content to let the reader decide.

# 1.

# The Reverend Dr John Polkinghorne, KBE

## Physicist

Those people who think that there is no room in science for a person of deep Christian faith – that the two world views are completely opposed – might have a problem with the Reverend Dr John Polkinghorne. He holds a PhD in quantum field theory, a DSc for research on theoretical elementary particle physics and is a Fellow of the Royal Society.

Raised in a Christian family, in 1979 – after eleven years as Professor of Mathematical Physics at Cambridge University – he resigned to study for the ministry. Since his ordination as an Anglican priest in 1982, he has written and taught extensively on the relationship between faith and science. He was appointed President of Queens' College, Cambridge in 1989 and was knighted in 1997. In 2002 he received the Templeton Prize – a prestigious global award made to the person who has done most to advance spiritual understanding.

*John, your expertise in science and theology testifies to the fact that for you faith and science cannot be mutually exclusive.*

*Does science, based as it is on rationalism, depend on scientists having some faith?*

Scientists certainly need to commit themselves to the belief that the physical world is intelligible and that there are helpful ways of thinking about it. If scientists tried to proceed by being sceptical about everything, we would never get anything done. A scientist will tend to believe the current theory until something comes along which makes them revise it. That, of course, is not dissimilar to the approach a religious person takes.

Faith is not a question of shutting your eyes, gritting your teeth and believing impossible things because some unquestionable authority says that's what you have to do, no questions asked. Faith is belief and commitment. It is a leap – not into the dark but – into the light.

*So science and faith are not in conflict?*

No, I see them as being complementary. Science is looking at one aspect of the world – telling us how things happen, what is the process of the world, how matter behaves and things like that. Religion is asking a deeper question – is there a meaning and purpose behind what is happening? Is there a divine will – a mind behind what is happening?

It seems to me that you need to ask and answer both questions. I need to have 'binocular' vision – to look to science and take very seriously all that science tells me about the world, and to look with the eye of religion and gain its perspective. I need to see the reality of creation and the Creator. There are two views which need to be fitted together – they have to be consonant with each other.

For example, I could tell you I intend improving my garden. When you ask how, I say by covering it in green concrete. The two ideas – gardening and concrete – don't really fit. So too, there is a dialogue between science and religion. Their two different perspectives have to fit together if they are perspectives on a single actual reality.

After Sir Isaac Newton tried to explain the world in terms of gravity and laws of motion, people would describe the world in terms of a clock – finely engineered, regular, ordered, mechanical, predictable. If creation was a clock, then the Creator was a clockmaker – piecing it together, winding it up and letting it go. For Newton, the world was mechanical and predictable. However, thinking has moved on.

*Your scientific speciality is quantum theory and subatomic particles. Could you give a quick layman's guide to QT, please?*

Quantum theory says that the world of atoms (and particles smaller than atoms) is very different from our everyday world. The quantum world is very cloudy – you can know where something is, but then you won't know what it is doing. Our everyday world seems very reliable but the quantum world is a world of probability rather than of predictable certainty. Scientists cannot be sure what atoms and particles will do. We may, for example, be able to say that half the atoms in a lump of uranium will decay in the next hour, but we cannot tell which individual atoms they will be.

Scientists believe that – even though they are largely unseen – atoms and subatomic particles are real. We believe they are real because they make sense of a great swathe of things we experience in the everyday world.

*Dealing as it does with a belief in an unseen reality, a belief in quantum theory doesn't seem dissimilar to the mechanics of religious faith. Is this the way you see it?*

I would certainly defend belief in the reality of an unseen God on exactly the same grounds as quantum theory. There is individual religious experience, the collective religious experience of worshipping communities and, of course, the foundational religious experiences of the Church and the New Testament. I then make sense of those experiences in terms of my belief in God the Father and the Lord Jesus Christ.

*Another scientific theory which requires an acceptance bordering on faith in the unseen is chaos theory. What is chaos theory about?*

Chaos theory says that the everyday world is not as predictable, tame and controllable as people once thought. Scientists talk not only in terms of clocks as they did after Newton, but also in terms of clouds. By this they mean there are systems in the world that are so exquisitely sensitive that any disturbance totally changes their future behaviour and the behaviour of dependent systems. Scientists describe it in terms such as, if a butterfly flutters its wings in Africa, the movement of air might multiply and, in a few weeks' time, could produce a storm thousands of miles away.

*What insight into religious faith can chaos theory offer?*

Chaos theory helps me see that things are interdependent and the future is open because the world is not clockwork. And it seems entirely reasonable to me for God to act in the world in a similar way. So I believe not in a God who simply set the world spinning and then sat back. I believe in a God who providentially interacts with history. It's a theory science and philosophy are still wrestling with.

*It is the belief that God intervenes in our personal histories that encourages people to pray. How do you think prayer works?*

Firstly, prayer is not magic. For example, if someone is seriously ill, people pray for that person's healing. Sometimes that person's wholeness comes through physical recovery but it may come through accepting the imminent destiny of death.

Nor is the world God's puppet theatre, where he's pulling all the strings and we have no say. Nor does God just sit back and let things happen. God interacts but he doesn't overrule the acts of creatures. Because he is a God of love, he has given creatures freedom to be themselves. So the unfolding history of the world is an interaction between the providential action of God and the individual acts of creatures, particularly creatures with free will like human beings.

Because I believe God acts in the world, it seems reasonable that we can ask God for things. If you believe that God is a personal God, then you would expect God to do particular things in particular circumstances. The question then is: Why do we need to pray? Why doesn't God just get on with it?

I believe we humans have some power to bring about the future and God has kept some power to bring about the future. When we pray, we commit ourselves to trying to align our will with God's will. When human and divine wills are aligned, things become possible which are not possible when they are at cross-purposes.

It's like laser light. Laser light is powerful because it is what physicists call 'coherent'. All its waves are in step – all the crests add up, all the troughs add down. That makes it very powerful. What we are looking for in prayer is a laser-light coherence between the human and divine wills. That will make things possible that would not be possible if we didn't offer ourselves to God in prayer.

*One of the answers to your own prayers was the sense that God was calling you to be an Anglican priest. How dramatic were the circumstances surrounding that time in your life?*

Not very. God speaks to different people in all sorts of different ways and it is very important not to make one's own experience the definitive model. I regarded becoming a scientist as a calling. I enjoyed being a physicist very much, but after twenty-five years I decided I'd done my bit for science. I talked with my wife Ruth and started to pray about what to do next. I imagined I'd retreat to a convent or somewhere, have a big crisis and all would be revealed. It didn't happen like that.

Because Christianity is central to my life, I started thinking about ordination. I chatted with some close friends and after a few months, both Ruth and I concluded that this was the right thing to do. It was undramatic but genuine guidance from God.

God often speaks in unspectacular ways to people, very often through circumstances – an opportunity comes along, something needs to be done. God deals with us as individuals. He doesn't have a one-size-fits-all approach. Of course, once we've heard what God wants us to do, we've got the painful necessity of following it through!

*You were brought up in a Christian family, how would you describe your personal coming to faith?*

I've had a sense of the Almighty for as long as I can remember. There have been moments of deeper Christian understanding and commitment but I have always been part of the worshipping community. Again, there have been no great dramas or crises.

I never felt I had to make a choice between faith and science – either one or the other. Today, as scientist-priest, I am missionary twice over. I try to tell the scientific community of the reasonableness of the Christian faith and I encourage them to talk about things other than their speciality. I raise questions like: Is the universe a series of happy accidents or is there a mind behind the order? Is there a purpose behind the fruitfulness? What do we make of Jesus? What makes him different from other founders of religions? What about the resurrection?

My message to Christians is: Don't be afraid of science. Christians are seeking to serve the God of truth. We should, therefore, welcome truth in whatever source it comes. It won't all come from science but some of it will.

*Do you think that science will ever present the world with a proof that God exists?*

No, I don't think you can prove the existence of God. In fact, I don't think you can even prove the existence of quantum things like electrons. It seems to me, you can have good reasons for believing in them but you can't prove them in a strict mathematical sense.

The line of thinking that helps us to reach well-motivated, justified beliefs in science is very similar to the line of thinking that enables us to reach well-motivated, justified beliefs in the sphere of religion. We can't prove God but we can experience him.

*John, do you believe that the Bible is the inspired word of God?*

Yes. I do. Along with most Christians, I believe the Bible is the inspired word of God, but not one which was divinely dictated. The Bible was written by human beings and it reflects the setting of its time. There are things in the Bible – for example, slavery – that are not models for Christian life today.

I think the Bible is a reliable record of Israel's history and encounter with God and of the life, death and resurrection of Jesus Christ, which is the centre of my faith. Without the Bible, we would know little about Israel or Jesus. The Bible is authoritative for me in the sense that it is a record of God's self-revelation. I want to read it carefully, because it is important to figure out what you are reading. For example, you shouldn't read poetry as if it were prose or you'll make some bad mistakes.

*How do you read the biblical account of creation in Genesis 1 and 2 – literally or as more of a word-picture?*

This is an instance where we need to figure out what type of literature we're reading. If we think Genesis 1 and 2 is a divinely dictated scientific textbook that God wrote to save us the trouble of having to do science, then we're making a big mistake. These chapters are theological writings.

The main message of Genesis 1 is that nothing exists except through the will of God. Fourteen times in Genesis 1 God says 'Let' as in 'Let there be . . .' We need also to read the 'book' of Nature, which God has also written. Science, too, tells us a lot that doesn't destroy the theological message of the Bible.

*How do you regard the Big Bang theory? Does it, for example, destroy the validity of the biblical account of creation?*

Big Bang cosmology is pretty well motivated. What happened very close to the Big Bang is a matter of controversy and speculation. The reason it is speculative is that it requires quantum theory. But because it is the universe you also require Einstein's general theory of relativity, and as yet scientists can't fit the two theories together consistently. So it is all guesswork – clever guesswork.

It is reasonable to suppose that the broad picture is right – that time, space and matter all came into being together. I see that as coming to being through God's creative word. The further back in time scientists go, the more the rules of science break down. In a sense, the Big Bang is a mystery. To me, creation is not about how things began, it is about why things exist. God is as much the Creator today as he was fourteen billion years ago. Big Bang cosmology is very interesting but I don't think that anything tremendous theologically hangs on it.

*Part of Big Bang thinking says that nothing, including time, existed before the Big Bang. Did God exist before the Big Bang?*

Yes, of course, he did. God is not in thrall to time in the way that we are. God exists eternally. God interacts with time. His most dramatic interaction with time is in the incarnation of Jesus Christ. But God is not bound to time.

*Some Christians believe that another scientific insight – the theory of evolution – conflicts with the creation accounts in the early chapters of Genesis. Where do you stand on evolution?*

Broadly speaking, I believe the evolutionary story. I think life started very simply. It has a long history and has become rich and complicated. That alone tells us that God chose to unfold creation patiently. Evolution is the story of how creation explores and brings to birth the very deep-seated fruitfulness with which its Creator has endowed it. God could have snapped the divine fingers and brought into being a ready-made world where everything is fixed and nothing changes. But God has done something cleverer than that. He's created a world that can make itself. That making of itself is called evolution.

*When things go wrong in his creation, people often blame the Creator. What insights can science offer into the problem of suffering?*

This is the most difficult of all questions. It is one which holds people back from faith and one which causes some believers to lose faith. There's no one-line answer. But I think science can give us an insight.

We have talked about an evolutionary world – one allowed by the Creator to make itself. It seems to me that this is a very great good. But the good has a cost because what drives evolution, what produces new forms of life, is genetic mutation. But if you have genetic mutation, some cells will be able to mutate and become malignant. You can't have one without the other. So, for example, the existence of cancer in the world is the shadow side of the fruitful history of the world.

We tend to think that if we had been in charge of creation we would have done it better. We would have kept all the nice things and got rid of all the bad things. But the more science understands the world, the more it seems a sort of package deal, that there is inevitably a good side and bad side to many things.

Such an insight suggests that at least some of the suffering of the world is not gratuitous. It is not something that could be avoided if only God took a little more trouble or cared a little bit more about people. One of the essential Christian insights into suffering is that God is not just a compassionate spectator, but that in the cross of Jesus Christ, God has participated from the inside in the suffering of the world. The Christian God is the crucified God as well as the redemptive God.

*Speaking of the Cross, can science explain the resurrection?*

No. The resurrection was a unique new act of God. It was not the mere resuscitation of a corpse. It was a transformative act involving continuity and discontinuity. Christ still had the nail prints of the crucifixion and he was able to appear and disappear at will. He is alive for evermore. The resurrection was a new act of creation. It shows that there is a destiny beyond death and that those who choose to do so will share in that life.

The resurrection marks Jesus out as different from any other religious leader. All the other great figures, like Moses and Muhammad, died in honourable old age, surrounded by their disciples who resolved to carry on their master's work. Jesus died in mid-life, deserted and with a cry of dereliction on his lips: 'My God, my God, why have you forsaken me?' If that were the end of the story, we would never have heard about him. It is the resurrection that makes the difference.

*What light can an understanding of genetics shed on the virgin birth?*

Again, there is no scientific explanation. The difficulty scientifically would be that a male child has to have a Y chromosome and a woman can only provide X chromosomes, so it is very difficult to see how that could happen. There are various extraordinary scenarios that might produce it but I think that they are theologically negligible.

The story of the virgin birth is seeking to say that Jesus' birth involved both God's initiative (it wasn't just a happy chance) and human collaboration (summed up by Mary's 'Be it unto me according to your word'). Whether you believe in the virgin birth or not depends on whether you think it is theologically credible. I believe that Christianity is about what one might call 'enacted symbol'. There is a power in symbolic stories and there is power in true stories. The whole Christian understanding of the incarnation, from birth to death, combines the power of history and the power of symbols – it is an account of enacted events.

It is possibly more difficult to believe in the virgin birth than the resurrection but I believe in the virgin birth on theological rather than scientific grounds.

*We've spoken much about origins. Let's talk about endings. How do you see the world ending?*

If things unfold as they are, the world will continue for a long time, hundreds of billions of years. But it will end, either in decay or collapse. If it gets hotter it will collapse with a bang. If the universe continues expanding it will cool and end in a whimper. Either way, the world is going to end in futility. Therefore the sort of evolutionary optimism that says the world is getting better and things will work out just won't happen. The universe is going to die just as certainly as you and I are going to die. Both sorts of life and death – cosmic and human – raise the same question about what is God up to.

The only ground for a belief in a destiny beyond either cosmic or human death is the faithfulness of God. That was exactly Jesus' argument when he described God as the God of the living, not of the dead. Because Christian hope depends upon God's faithfulness, we can have a sure and certain hope.

## FACT FILE

*Reverend Dr John Polkinghorne, KBE*
MA (Mathematics), Cambridge.
PhD in Quantum Field Theory, Cambridge.
ScD for research on Theoretical Elementary Particle Physics, Cambridge.
Knighted 1997.
Fellow of the Royal Society.
Fellow of Queens' College, Cambridge.
Canon Theologian, Liverpool Cathedral 1994–2005.
Lecturer in Mathematical Physics, Edinburgh 1956–1958.
Lecturer in Mathematical Physics, Cambridge 1958–1965.
Reader in Mathematical Physics, Cambridge 1965–1968.
Professor in Mathematical Physics, Cambridge 1968–1979.
Trained for Anglican priesthood 1979.
Ordained deacon in 1981 and priest in 1982.
Parish appointments 1981–86.
Dean and Chaplain, Trinity Hall, Cambridge 1986–1989.
President, Queens' College, Cambridge 1989–1996.
Winner of Templeton Prize for Science and Religion 2002.

Founding President of the International Society for Science and Religion 2002.

Author of more than thirty books.

2

# The Reverend Professor Alister McGrath

## Biophysicist

Born and raised in the churchgoing Northern Ireland of the 1950s, Alister McGrath was educated at Belfast's Methodist College. Today he is a prominent writer on Christian thought and an outspoken critic of atheist scientist Richard Dawkins. He is the director of the Oxford Centre for Christian Apologetics and Professor of Historical Theology at Oxford University.

It is easy to imagine that his journey from Christian background to Christian spotlight was straightforward. Easy, but wrong. He describes his teenage years as a time he was a 'Marxist atheist', a label he wore when he went to Oxford to read chemistry. He emerged from his studies with a first-class honours degree in chemistry (which was followed by a doctorate in molecular biophysics) and a vibrant Christian faith. He later gained a first in theology, studied for the Anglican priesthood, was ordained and subsequently became a (non-honorary) doctor of divinity at Oxford. Alister doesn't believe that being a Christian means having to commit intellectual suicide.

*Alister, how would you describe your journey from atheism to faith?*

As a child I never had any interest in Christianity. I went through the motions of going to church with my parents but neither my heart nor my head was in it. It was while I was at the Methodist College, probably aged around fifteen or sixteen, that I became an atheist – somebody who deliberately and intentionally does not believe in God and thinks that anyone who does believe in God is mentally deficient or seriously screwed up.

Growing up amid sectarian violence, I concluded that if there were no religion, there wouldn't be any violence. At the time Marxism, with its offer of political transformation and its aim of doing away with religion, was very attractive to me. I was also studying natural sciences. To me, science had disproved God. So I was a Marxist atheist who enjoyed sciences. I went to university as a convinced atheist but then I started to have one or two questions, such as: 'If atheism was right, then why were so many people religious believers?' I told myself it was because they were fools. Part of me was very happy with that answer but part of me knew that actually it wasn't good enough. Then I discovered some very articulate Christians at the university. They showed me that atheism was not as robust as I had thought.

Gradually I began to realise that I'd misunderstood what Christianity was. I had thought that it was simply a kind of ritualistic, mechanistic thing, all about keeping rules. I had no idea that it was really about a personal relationship with Christ. Discovering that changed things in a very big way. I discovered not simply that Christianity was true, but also that it was real. It was not just something that made sense but also something that could transform someone's life. I decided I wanted to become a Christian. I can't point to a single defining moment but when I went to Oxford in the October I was an atheist; when I went home for Christmas I was a Christian.

*As a scientist-theologian, you span the intellectual divide between two disciplines which are often seen to be in conflict. Are science and religion, specifically Christianity, mutually excluding?*

No, they're not. In fact, many Christians who are scientists would say they find that these two disciplines reinforce each other. There's a very strong religious motivation for scientific research. If you believe that God made the world, you can get additional insights into God by studying nature. Science can't answer the big questions such as 'Why are we here?' or 'What's life all about?' In many respects, science is raising these questions but not answering them. Therefore it's essential for a theological narrative of things to run alongside the scientific account. Science and Christianity reinforce and complement each other.

*So science is not necessarily atheistic*

Science is neutral. It does not presuppose or imply atheism. Darwin's theory of evolution, for example, is not by definition atheistic. Some people just interpret it that way. A basic assumption in science is that we do not presume God. But science does not in principle exclude God. Science simply says we are not going to bring God into things as a matter of principle. Therefore someone can interpret the sciences in an atheistic way or a Christian way. Richard Dawkins asserts very strongly that to buy into modern science is to say that there is no God. But that is simply not true – scientifically or philosophically. There are a large number of Christians who are scientists who spend most of their professional careers disproving him on this point.

*You've had evolutionary scientist Richard Dawkins under the microscope for a number of years. In 2004 you published Dawkins' God, in which (in the style of gamekeeper-turned-poacher) you attacked Dawkins's atheistic worldview. In 2007* The Dawkins Delusion? *was your response to Dawkins's best-selling* The God Delusion. *What makes you want to challenge your Oxford colleague?*

Dawkins scores points by misrepresentation – by presenting arguments in their worst possible light and by choosing extreme Christians as though they are representative of mainstream Christianity. He works on the assumption that his readers know very little about Christianity. He argues, for example, that God is a child abuser who encourages infanticide. He asserts that if you believe in evolution, you cannot believe in God, because evolution is by definition atheistic. But that is a very inaccurate interpretation.

Dawkins also interprets a Christian's 'faith' as 'blind trust'. To him 'faith' means 'running away from evidence'. But that's not a Christian definition of faith. Christians will say that faith is about believing in a God who not only exists but who may also be relied upon utterly, someone into whose hands I can entrust myself, knowing that he's going to guard me and keep me. People like simple answers to hard questions. That's why Dawkins is so popular. When I was an atheist, I sounded like Richard Dawkins. I focused only on the things that fitted my theory. One of the things that made me stop being an atheist was realising that things are rather more complicated.

*Science, then, does not have all the answers. So what insights can Christianity give to seemingly insoluble questions, such as: Why does God allow suffering?*

I don't think we're ever going to explain suffering completely. Christianity doesn't offer us a neat theory of suffering. It speaks to us of a God who is present in suffering. In other words, when we suffer, we don't suffer on our own. Psalm 23, for example, says that even in the valley of the shadow of death God is with me. How did God redeem the world? Through suffering – the suffering of Christ, his Son. It was a suffering that one day will bring about the complete elimination of suffering from the world.

There are no easy explanations but there is this reassurance of a God who has been through suffering and will be with us as we suffer. The fact that we find suffering so distressing is actually saying something very significant, as if there is something built within us that says: 'This isn't right.' I believe this instinct comes from God. And it's saying that things won't be like this for ever, one day it will be changed. There is more to life than the physical world and God urges us to find it.

*Based as it is on rational observation and explanation, what can science add to the idea of miracle?*

It is commonly said that science disproves miracles. It doesn't. Science says that there are certain things that by their nature are very improbable. So improbable, in fact, they shouldn't happen at all. Science can't say things can't happen, merely that they are very improbable. For the Christian, miracles seem to have this tendency to happen around Jesus. For the Christian, therefore, modern science makes these events all the more remarkable and it forces us to ask the question: 'What is so significant about this man who does all these things?' Science does not disprove the resurrection of Jesus. It says that the resurrection cannot be explained by a natural process. The question is not so much 'How did the resurrection happen?' but 'What does it mean?'

*Physicist Sir Isaac Newton pictured God as a clockmaker, with the world running strictly to predetermined patterns.*

*His cause-and-effect physics has been enhanced by chaos, quantum and relativity theories. How do you see God? A clockmaker, winding creation up and letting it go? A puppeteer, pulling the strings and putting words in people's mouths?*

I don't believe we are simply puppets whom God manipulates. Nor is God someone who leaves us to our own devices. God is one who guides. He shows us the path to take but expects us to make the decision whether or not to take it. One of the most astonishing things about the Christian gospel is that even when we take a wrong path, God is still able to use us. God wants the best for us, and tries to guide us back on to the right track, but he doesn't impose on us – he doesn't deny us our responsibility.

*One area of contemporary scientific thought is the quest to explain life in one formula – the theory of everything. Will that ever happen or is it a scientific holy grail?*

Stephen Hawking says it is unattainable. But I think it's worth attaining because if you believe God made the world, it means that there is some intrinsic rationality to the world which reflects the wisdom and the justice of God. The theory of everything says there is one big thing that explains everything – one place where the buck stops. But that's exactly what Christians have been saying about God: the buck stops with him. God can explain but does not need to be explained. The theory of everything – the method of making sense of the world, the theories that we know – could simply be an explication of the mind of God.

*When it comes to Christian ideas about how the world started, there are a number of models. Could you explain them, please?*

All Christians take the Bible very seriously and say that it is of the utmost importance to remain faithful to what it is saying. The key question is how to interpret the Bible. Are, for example, the first three chapters in Genesis literal history? Or are they something deeper than that?

There are four main positions within Christianity. Young Earth creationism says the Earth is about six thousand years old and was made pretty much in the form that we see now. This conclusion, they say, is a natural reading of the Book of Genesis. Another school of thought – old Earth creationism – while agreeing that God made the Earth and everything in it within a limited time frame, says that there are gaps within the Genesis account. For example, God creates the universe, then there's a gap and then he creates something else. In other words, the universe is extremely old but God made the Earth and us at a much later stage. That, again, is a perfectly legitimate way of interpreting the Book of Genesis.

A more recent development is Intelligent Design. This is quite similar to old Earth creationism but the talk is not so much about 'God' as about an 'Intelligent Designer'. The basic argument is that we cannot explain the way the world is by purely naturalist explanations. We have to invoke an Intelligent Designer to explain what we find. This argues that a purely Darwinian account of the world cannot give a total picture.

The fourth approach is theistic evolution. This sees evolution as the way in which God providentially exercises his creative processes and brings the world into being. This approach finds a lot of favour among Christian biologists and links in well with some parts of the Book of Genesis, particularly those that talk about the Earth bringing forth things, which seems to imply there's some kind of ongoing natural process.

Each of these ideas has its strengths and its weaknesses. But underlying each of these approaches is the very firm insistence that the world is God's creation; it is his world not ours; it was deliberately brought into being – and so were we.

*So it is possible for somebody to be a Christian and to believe in evolution?*

Yes, it is. Evolution is not, by definition, atheistic. Darwin saw his theory as reconcilable with the Bible. He struggled with his Christian faith towards the end of his life but that was because his daughter had died very young, not because of his ideas on evolution. Some Christians will be uncomfortable with the idea of believing in evolution, particularly because it raises the question of how to interpret the early chapters of Genesis. That's a very big issue in its own right. All I can say is that, with complete integrity, there are many Christians who see evolution as illuminating the way in which we understand Genesis and as giving us an enhanced vision of how God brought the world and humankind into being. People can make evolution atheistic but it doesn't have to be.

*Alister, why is it important to you that God created the universe?*

Well, firstly, it is saying that neither the universe nor humankind is an accident. We are meant to be here. It affirms the idea of purpose. It also raises the question: Why are we here?

Secondly, a God-created universe tells us that the world itself is not divine. In the ancient world, people thought the Sun, Moon and stars were divine and had to be placated, otherwise there would be trouble. But the doctrine of creation says that if you know God and you're right with him, then ultimately there's nothing to fear. We don't have to worry about occult forces.

Thirdly, creation is a beautiful reflection of the fact that the God who made everything we can see also made us and cares for us. In Christian thinking, however, the world as we see it is not the world that God made. It is a fallen world. The created order is showing suffering and pain. Christians believe that at the end of all things God will bring about a place where there is no more suffering, no more pain, no more death. In other words, Eden will be restored. This inspires Christians to bring about those conditions now. That is why, over the years, so many Christians have worked towards the alleviation of suffering, illness and poverty. And why we still do. The Christian doctrine of creation says we've lost our way but it is redeemable. Which is, of course, why Christ came.

*Do you believe that the creation account of Genesis 1 and 2 is literally true?*

The idea that God created the world is literally true. That belief is not only in the opening chapters of Genesis but throughout the whole Bible. In the opening chapters of Genesis we find a particular way of thinking about how creation took place. It is trying to set before us, in a way that is true but not necessarily true at the absolutely literal level, how God did this. We are asked to try to make sense of this, bearing in mind that there is not simply this passage on its own but a whole range of other passages that are trying to help us understand what it means to say that God created the Earth, and God created us.

*Would you say that the Bible is the inspired word of God, and if so, what do you mean?*

Yes, it is. By this I mean that the Bible has an origin and an authority which no other human text has. I also mean that, as I read the Bible, the same God who was responsible for this text will illuminate my mind as to its meaning and how to put into practice what it says. It is said that the Bible is like a wax nose – you can twist it to mean what you like. Talking about the inspiration and authority of the Bible doesn't lead us to the conclusion that there's only one way of interpreting it. That's why it's important for Christians to test what we conclude from our reading with other people and with the Christian creeds.

*Did Adam and Eve exist?*

I've no doubt that there was an Adam and an Eve in the sense of some primal figures. Throughout the Bible Adam and Eve also are often used as representatives of the human race. The Bible includes a number of different literary forms, including history, prophecy, allegory and poetry. Each needs to be identified and interpreted in an appropriate way. The difficulty is trying to decide which category a particular biblical passage belongs to. If, say, we were to take poetry as literal truth, when the poet writes about 'mountains dancing for joy', we might end up in some difficulty. It is important to appreciate each level of meaning, really to try to come away from reading the Bible with an enhanced sense of its richness, rather than simply saying every bit has to be interpreted in exactly the same way.

*Can somebody be a Christian and believe that the universe came into existence through Big Bang?*

Big Bang theory is a good approach for Christians to adopt. Early in the twentieth century the prevailing wisdom in the scientific community was the steady state theory. In other words, the universe has always been here and while it may fade away at the edges it is being replenished from the centre. Big Bang changed all that. Interestingly, many atheist scientists were strongly resistant to Big Bang not on scientific grounds but because they thought it sounded too religious.

The key point here is that Big Bang is a major scientific advance which seems to reinforce what Christians have always been saying – God created. For many Christians this is simply a scientific version of the Christian doctrine of creation. Big Bang doesn't explain everything but it raises some very big questions, including, put crudely: 'Who pressed the button?' There's an easy Christian answer to that. It can be explained in very complicated ways but the basic idea is that God was there right from the beginning. Just as it says in the Book of Genesis!

## FACT FILE

*The Reverend Professor Alister McGrath*
BA (Chemistry), Oxford.
DPhil (Biophysics), Oxford.
BA (Theology), Oxford.
BD (Theology), Oxford.
DD, Oxford.
Ordained as an Anglican deacon in 1980 and priest in 1981.
Curacy in Wollaton, Nottingham 1980–1983.
Lecturer in Christian Doctrine and Ethics, Wycliffe Hall, Oxford 1983–1993.
Research Lecturer in Theology, Oxford 1993–1995.
Principal, Wycliffe Hall, Oxford 1995–2004.
Professor of Historical Theology, Oxford 1999–present.
Director of the Oxford Centre for Christian Apologetics 2004–present.
A former atheist, Alister McGrath now debates with leading modern atheists, including Richard Dawkins and Christopher Hitchens.
Author of more than twenty books, his most recent book is the international bestseller *The Dawkins Delusion?* (with Joanna Collicutt McGrath, London: SPCK, 2007).

# 3

# Professor Frank Russell Stannard, OBE

## Physicist

He knows how nuclear bombs work. He believes there could be life in other galaxies. He reckons science doesn't have all the answers and calculates cricket to be the best thing since split atoms. His love for both started while he was a pupil at the school which overlooks the Oval cricket ground. With E equalling mc-square leg, Surrey CCC fan Russell Stannard went on to read physics at London University. What he discovered – the work of Albert Einstein – hit him for six. He became a high-energy nuclear physicist. For twenty-six years Russell was Professor of Physics at the Open University. In 1989 he published *The Time and Space of Uncle Albert*, a book in which he explains the wonders of Einsteinian physics to children. Rocket science it is, and two other Uncle Albert adventures followed, along with books on science and religion for adults and children. Now in retirement, he is as inquisitive and adventurous as ever. When he's not relating relativity theory to schoolchildren, he's shaping abstract sculptures in his garden workshop.

*Russell, Sir Isaac Newton's career took off after the fall of an apple, yours seems to have begun with the fall of a wicket. How did you get interested in science?*

I was a pupil at Archbishop Tenison's School next to the Oval. I was mad keen on cricket and there were only two sets of windows that overlooked the ground – the staff room and the physics lab. So the only way I could keep an eye on my heroes was to do physics. Then at university I came across Einstein's theory of relativity. It completely blew my mind.

*Could you give a swift analysis, please?*

Relativity is all to do with what happens when you go very fast. You can't travel faster than the speed of light because the faster you go, the heavier you get, which makes it harder to go any faster. Einstein said that time slows down the faster you go, so if you could go fast enough you could live forever.

It was the most wonderful thing I'd ever heard. I knew right away that I must spread the word that the world in which we live is much more interesting than we would gather from the sorts of science lessons we had at school. Most of the people I talked to didn't believe me: How can you live forever? I went on to do some experiments on subatomic particles, where we make things go at about the speed of light and time does slow down. I saw it for myself and knew I must get relativity across to unprejudiced minds – to children.

*Why do things get heavier the faster they travel?*

They're gaining energy and energy has mass – heaviness. Even stationary things – a coffee cup, for example – have mass. It comes from a locked-up kind of energy. In fact, a cup has enough locked-up energy to blow up a town, if it ever got out. A coffee cup is safe because its materials are very stable but uranium, for example, is very unstable. By rearranging the material, the nuclei, you can release some of the locked-up energy. Just a tiny fraction is enough to create a nuclear bomb. The energy of a nuclear bomb comes from a matter that has been transformed into another form of energy – heat and light. And that affects us all.

*Where did your Christian faith begin?*

As a child I seldom went to church. My first sort of religious awakening came when I was in the sixth form. As school captain, I had to read the lesson in our annual Founder's Day service in St Martin-in-the-Fields. It just had an effect on me. I can't really describe it. I started going to St Martin's, occasionally at first and then every Sunday. I felt at home there, I felt I belonged, that this was a natural place for me to be. I was confirmed when I was about nineteen. I can't remember a thing from the confirmation lessons, but there was something about the curate who took those lessons, Austin Williams, that impressed me. For the first time in my life I was in the presence of a man of God and I wanted to be like him. He had something I wanted. It was a classic case of religion being caught rather than taught.

In 1966 I met another vicar – Father Nad, who was a converted Hindu. He suggested that I ought to become a reader in the Church of England. There was no conversion point – no one moment when I was born again. It was more of a process. I'd be preaching on a Sunday and be in the laboratory Monday to Friday. To many people, the two didn't go together. How could you be a Christian and a scientist? One of my responses was to write a book, *Science and the Renewal of Belief*.

*What would you say your faith gives you?*

My faith gives me strength in times of trouble. Purpose. Vision. It makes sense of the whole of my life and gives me hope for the future. To me, believing in God just makes good sense. I can make more sense of my life in the context of believing in God, than having to account for my life if I assume there isn't a God.

*To what extent can science prove the existence of God?*

Many people are searching for something spiritual but feel they cannot look to traditional religion because it has been caught out by science in the past. They think, for example, that Christianity has been shown to be untenable for any intelligent, well-informed person. And if they are to have a spiritual life, they want one with intellectual integrity. So if 'becoming religious' entails accepting the Adam and Eve story literally and turning one's back on evolution, Big Bang and stuff like that, they feel they cannot make a stand because it would be living a lie.

Science is not an obstacle to religious belief. Much of science is as irrelevant to religious belief as it is irrelevant to the likes of music or poetry. Science cannot, for example, account for the resurrection. Science supports religion but not in the sense that you look to science for proof of God. There are interpretations of the Bible which are completely consistent with modern science.

Once you embrace the findings of science – as scientists reveal more about God's world, the same God that you encounter in your prayer life – then you start to see an enormous amount of enrichment coming into your understanding. Nobody ever gets argued into a loving relationship with God. Science neither proves nor disproves his existence. The strongest evidence for God comes from your own experience, what you get out of your relationship with him. That is something a person has to try for themselves. Unless you have honestly tried to pray, to enter into that relationship and sense the presence of God then arguing is a waste of time.

*How would you describe your style of faith?*

I am an orthodox (with a small 'o') Christian. I believe in the resurrection of Christ. I believe in life beyond death. I see great value in the doctrine of the Trinity. I believe Jesus was fully man and fully God. Perhaps, as a scientist, it's easier to believe these two states can coexist. After all, Einstein once showed how, under certain conditions, a particle can be both confined to a point and at the same time be a spread-out wave.

I have difficulty with accepting the literal truth of the virgin birth, mainly because accounts of it didn't come into being until many years after the birth of Christ. The earliest writings we have are Paul's letters. In them Paul makes great arguments that Jesus was someone absolutely special, that he was the Son of God. I find it curious that Paul didn't support his argument by mentioning the virgin birth as evidence. However, I still find great value in the story because it contains a very deep spiritual truth – Jesus was truly God and truly man.

I also believe God knows the future. I have no difficulty with this idea because it is part of relativity theory – we recognise that space and time are much more similar to each other than we had originally thought. People used to think of three-dimensional space as being one thing, and one-dimensional time, going from the past to the future, as something different. They believed that in space things moved along a time axis. Einstein showed us that this is completely wrong. Time is the fourth dimension. God, therefore, sees time – all of past, present and future – along with all of space as one block. Again because of science, it is easier for me to understand the future as something that is out there waiting for me to come across it, and that God already knows it.

I believe in Heaven but I've no idea what it'll be like. I hope God's got something better in mind than what I can dream up. My tongue-in-cheek view would be to get to the Pearly Gates and St Peter says: 'Thank goodness you've come, Russ. God's got a lot of jobs for you to do and you are already behind schedule!' That would be wonderful. I hate being bored.

*So would boredom be your idea of Hell?*

I don't see Hell as being a deliberate place of torture. Everything we know about God points to the fact that he is not vindictive. Hell is unbelievers being in Heaven and simply not understanding why the others are having such a good time. I once went to a Wagner opera with a friend. He thought Wagner was the business. For three hours he sat there in raptures. He was gone. Me, I was bored out of my mind. I couldn't get on the same wavelength. For him it was Heaven, for me it was Hell.

*What about the heavens, black holes and all? In your Uncle Albert stories, you have Einstein calculating that the universe contains one hundred thousand million galaxies each with one hundred thousand million stars. Is there life out there?*

We don't know. The only way we're going to find out is by making contact. The SETI – Search for Extra Terrestrial Intelligence – project is trying to do that right now. The fact that we've found nothing yet doesn't mean there's nothing there. There are probably billions and billions of Earth-like planets out there. But is there intelligent life? I tend to go along with the idea that we are nothing special, that it is arrogant to think that the universe was made on that scale just for us here on Earth. So yes, I would say there probably are many examples of forms of life, at least as intelligent as humankind. They won't necessarily look like human beings. There might, for example, be great survival benefits in having an eye at the back of your head. Or, if you want to survive buffet parties, in having three hands. And two long legs would be useful for the odd game of intergalactic cricket!

*Staying with planet Earth, though, how do you read the biblical account of Adam and Eve – as true or as truth?*

The Adam and Eve story tells us important spiritual truths about ourselves. Perhaps the most important of which is that basically people are selfish. Self-centred. Greedy. We are disobedient to God's will. We want to do things our way rather than his. Adam and Eve have access to all the fruit trees apart from one. And yet that is the one they want to go for. Why? Because they think they are the only ones who matter. Because of their disobedience, and because we are their sons and daughters, humankind is tainted with the same brush. From the moment of conception, we have this tendency to be sinful. What the Adam and Eve story shows us is that no matter how hard you try to cocoon a baby away from evil influences, that child will be selfish and will rebel against God – will sin.

*What insight can science give us into the concept of sin and its origin?*

Until the theory of evolution came along, the doctrine of original sin was something people had to accept on trust. Evolution tells us that humans are evolved animals. Our genes contain DNA coding which, in common with other evolved species, describes our physical characteristics. All these characteristics have been honed over the years in the struggle for survival. In the animal world it's the sharpest clawed and fastest runners that survive. In the case of humans, the survivors are those with the greatest intelligence. But also coded into the DNA are certain behaviour characteristics. There is, for instance, no point in having sharp claws if you don't know what to do with them. So an animal which is programmed to strike out at a prey and spike with a claw is the one that is going to get the meal, survive and pass on to its offspring that predisposition to strike without thinking. As soon as we recognise ourselves as being an evolved animal, then we've got to expect that encoded in our DNA will be certain behaviour characteristics which were conducive to the survival of our ancestors. Those in the main would be selfish, self-centred behaviour patterns like grabbing what food and shelter are available.

*But aren't humans more than just highly selectively programmed animals?*

There is an important difference between ourselves and other animals. Because of our intelligence, we have the ability to reflect on what we are doing. We have self-awareness. We can choose to act differently from our basic instincts if we feel there is sufficient reason. Animals are genetically determined. We are genetically influenced. Unless you consciously decide, 'No, I am not going to do that,' you will be selfish. (It is the legacy of original sin.) But we can consciously decide to be unselfish.

*So evolution and Adam and Eve are complementary rather than competitors?*

The Adam and Eve story was never meant to be a scientific account of our physical origins. It is purely concerned with timeless spiritual truths like, for example, Eve being made from the rib taken out of Adam's side. All that means is that man is not complete without woman and woman is not complete without man. It is talking about marriage. It is not talking about how women physically came into being. And you use either the scientific accounts or the Adam and Eve story, depending on the question.

Science is not in competition with faith. Science helps us understand that the Adam and Eve story got it right about our basic human nature. This is very exciting because it then opens up the possibility for one's scientific understanding to enrich one's religious belief, and that is how I see the relationship. But you have to start off by realising there's nothing contradictory between science and religion. And you have to start taking religion seriously.

*What about the creation of the universe? Does believing in the Big Bang mean there isn't a Creator God? Or is there no contest because it was God who caused the Big Bang?*

It is more complicated than that, because in scientific terms the Big Bang is a very special kind of explosion. At first we might think it is like the biggest explosion that has ever happened in the world; that it went off at a particular point in space and at a particular point in time. And if you're lucky you might get into a spacecraft and go off into outer space and eventually come across a blue plaque that says: THE BIG BANG HAPPENED HERE. It's not like that at all because the Big Bang marked not only the coming into existence of the contents of the universe, but also the coming into existence of space and the coming into existence of time.

Before the Big Bang there was no time, no space. In fact, correctly speaking, we can't even use the phrase: 'Before the Big Bang', because that supposes a time before time. This changes one's whole attitude to the Big Bang. Cause is followed by effect. Now in this context the Big Bang is the effect. So what caused the Big Bang? But there's a problem. The cause must happen before the Big Bang and there is no time to accommodate a cause!

*How does this help our view of a God who created the universe?*

Well, it gets rid of a very commonly held view that God, who has existed through all time, at some point decides to create humans and somewhere for them to live by lighting a blue touch paper and boom, we're on our way. That idea has to be scrubbed. Scientifically, there is no time before the Big Bang so there is no God before the Big Bang. It doesn't make sense. Linguistically it seems to make sense to think of God existing before the Big Bang but scientifically it doesn't make sense.

So does that get rid of a Creator God? No, because what one has to do is make a very clear distinction between two words, which in normal everyday conversation we muddle up. Those two words are 'origins' and 'creation'. If we're thinking about origins then we're asking how did something originate. If we're asking how did the universe originate then go to a scientist and he'll talk about Big Bang. If on the other hand we're talking about creation, we have totally different questions in mind: Why is there something rather than nothing? Why are we here now? What is responsible for our existence, what is keeping us in existence? And the religious answer is the Ground of All Being – that which we call God. And, as such, God sustains us through time. This is why when theologians talk about God the Creator they normally couple it with the idea of God the Sustainer, because God's creativity is something that is required throughout time. I see the world as past, present and future and that God is upholding it all equally at all instances of time.

*For some people the Adam and Eve story says that God hand crafted humankind, while evolution says that we're descended from lower life forms. Are we hand crafted or evolved?*

I believe we are descended from the same ancestors as apes. Ancestors which we would regard as more ape-like than ourselves. You have the original ape-like ancestors which then split up and become chimpanzees and baboons and humans so we are just one of the branches if you like. All the branches eventually go back to primordial slime. The religious question is: Where in this continuous evolutionary chain does the spirit come in? Traditionally one has always drawn a distinct line between humans and animals – that humans have a spirit and animals don't. Humans can go to Heaven and animals can't because they don't have a spirit. In the light of evolution I think the distinction is more blurred than that, and that just as we have had an evolution of our physical selves, we perhaps ought to be thinking in terms of a parallel evolution of our spiritual selves. By this I mean that our very primitive ancestors were concerned with little more than basic survival – sex, food, shelter.

At some stage one of our ancestors reached the level where they had sufficient thinking power to go beyond that and ask, 'Is there anything more than what I am naturally doing and seeing at the moment? Ought I to be doing this and is there some purpose to life?' As soon as that first ancestor started to have those sorts of thoughts, we get for the first time just a glimmer of a possibility of a rudimentary relationship with God. Today our understanding of the Almighty has grown immensely and much of that understanding is thanks to science.

## FACT FILE

*Professor Frank Russell Stannard, OBE*
BSc (Special Physics), University College London.
PhD in Cosmic Ray Physics, University College London.
Appointed OBE in 1998 'for services to physics, to the popularisation of science and to the Open University'.
Fellow of University College London.
Fellow of the Institute of Physics.
Lecturer, University College London 1960–1969.

Reader, Open University 1969–1971.
Professor of Physics, Open University 1971–1998.
Emeritus Professor of Physics, Open University 1999–present.
Licensed Lay Minister (Reader) in the Church of England 1966–present.
President of Science and Religion Forum 2000–present.
Author of more than sixty papers.
Author of fifteen children's books, including the *Uncle Albert* series on the science of Einstein, and of eight books for adults.
Radio and television broadcaster.

# 4

# Michael Poole

## Physicist

Biologists are not the only ones to discover that from little acorns mighty oaks do grow. Physicist Michael Poole's passion for science began when he was a lad at primary school. A classmate brought a battery and a bulb to school, joined them together with a piece of wire and the bulb lit up outside the torch. Michael was hooked. It was not the only light to go on. Michael teamed up with a friend and started performing experiments. Years later, he read physics at King's College in the University of London and then taught in a London comprehensive school before returning to King's College as a Lecturer in Science Education. Today he is a Visiting Research Fellow in Science and Religion at King's.

*Michael, would you describe, please, your personal journey to faith?*

I grew up in a churchgoing family but hadn't made a personal commitment myself. When I got to university I gave up going to church – to me being a Christian was for old ladies and young children. Then in my second year I met a really pleasant student who didn't fit with my 'old-ladies-and-little-children' view of a Christian – he was a medical student and played rugby. He kept trying to persuade me to go along to church with him. For ages I didn't want to know. But eventually I went to church. I was starting to wonder whether there was anything more to life than learning and teaching physics. This student said: 'If you want to know if God exists, why don't you start reading the Bible and praying – because these are ways in which God speaks to us.' So I did.

One evening walking around Streatham Common I came to faith. It was not anything spectacular, no blinding light, but that night I wrote in my diary: 'I am now a Christian. God showed me the way on Streatham Common tonight.' The following morning I was doing a vacation job in a nucleonics laboratory and I opened my diary. I went hot and cold when I read what I had written the night before. It wasn't like me; but I knew it was true. That was the beginning of my faith. I didn't have a lot of understanding. I understood my need of a Saviour and that my sins were forgiven. But that was just the beginning.

*For much of your career you have studied the interaction of the twin driving forces of your life: science and Christian faith. In the past, through the likes of the Galileo and Darwin affairs, science and faith have sometimes been portrayed as competitors. Today it is popular to write off faith as having nothing to contribute – Richard Dawkins speaks of a 'God delusion' and faith as a 'mind virus'. Many non-scientists look to science for all the answers. Does it have to be a case of either faith or science?*

No. I like the idea used by Francis Bacon and many others, who said that God speaks to us through two books: the book of Scripture – God's words – and the book of Nature – God's works. (Darwin also used this metaphor in his preface to *The Origin of Species*.) Galileo said that he didn't expect to see a contradiction between the two books, so long as we understood them both.

There are a number of areas of common misunderstanding about faith and science. The first is that faith and science are totally separate ways of interpreting the world. Some people think that science has all the facts and requires no faith, and that religion – particularly Christianity – is all about faith with no facts. This is nonsense.

Science makes use of reason but it does not have a monopoly on it. In the Old Testament, for instance, God invites his people: 'Come, let us reason together.' Science requires faith – confidence, trust. Scientists, for example, have to believe in the uniformity and orderliness of nature – that there are expected patterns of behaviour – otherwise they couldn't encode that orderliness in scientific laws.

Science requires faith that an experiment that was carried out yesterday will produce much the same results tomorrow. Perhaps the biggest faith investment for scientists is to believe that the universe itself is open to being understood. Einstein said that 'The most incomprehensible thing about the universe is that it is comprehensible.' It can be understood by humankind. Christianity is based on fact. The life and death of Jesus are well-recorded facts, placed within the time that the Romans occupied Israel. The resurrection of Jesus is the central fact and books like *Who Moved the Stone?* have been written on the evidence for it.

The second area of misunderstanding is that either faith or science can provide all the answers. There are a lot of unanswered questions in them both. If one is thinking about a God who created this universe with its one hundred thousand million stars in one galaxy among one hundred thousand million similar galaxies, it's not surprising that our comprehension is limited.

A third misconception is that scientific and religious explanations are interchangeable. Many years ago theologians were concerned that the growing success of science seemed to be displacing God. For example, people thought it was God who kept the planets in their orbits, then scientists said it was gravity. Feeling threatened, the theologians pointed to things that scientists couldn't explain and said: 'Ah, that's God.' Professor C.A. Coulson termed this as the 'God of the gaps'. Trying to fit God – an explanation in terms of agency and purpose – into a gap that should rightly be filled with science, mixes the two different types of explanations and doesn't make sense. Coulson, writing from a Christian perspective said that scientific ignorance shouldn't be explained away as God but should spur people on to become better scientists. God, he pointed out is involved in everything; both what we know and what we don't know.

A fourth misleading view is that science and religion are conflicting ways of looking at life. In the nineteenth century, many clerics were also scientists but there was a growing sense of wanting to make science a distinct profession rather than a hobby. Biologist T.H. Huxley had difficulties getting a job because he wasn't a member of the Church of England and so couldn't get into Cambridge or Oxford. Like Huxley, others also thought that science should break away from the Church, and they began to promote science as something which replaced religion as the main cultural activity. In 1860, Huxley and Bishop Samuel Wilberforce took part in a public debate about Darwin's theories of evolution. Although it was not widely reported at the time, some twenty years later the confrontation was resurrected and rewritten to promote the idea that science and religion are in conflict.

*One current worldview – notably put forward by Richard Dawkins, with whom you have debated in print – is that life can be explained without reference to God, a Creator, a Supreme Being, a First Cause. Is, then, science atheistic?*

Why should science be atheistic? Science is concerned only with the natural world, with things like energy, space and time. Religion asks questions such as: Is there anything other than the natural world to which the natural world owes its existence? Science can't answer that question. It asks different kinds of questions. I believe in rational scientific approaches but they are not alternatives to trust in God. There is nothing atheistic about science. We can have different types of explanations; the scientific ones are not the only ones. There are also explanations of agency and purpose. The idea that science is atheistic gives science an undeservedly bad name.

*If science and religion share an overlap of fact and faith, can science give insight – proof, even – that supports the two most unusual events associated with Jesus – the virgin birth and the resurrection?*

The resurrection is not a scientific question. It is an historical one. There is evidence, though. A group of frightened, disillusioned followers suddenly became a mighty power that imperial Rome had to reckon with. The testimony of non-Christian historians such as Tacitus, Pliny and Josephus all lend support to the story that something dramatic happened. There are the accounts in the Gospels and the personal testimonies of millions of Christians.

Science cannot explain the virgin birth. Science deals only with the normal patterns of behaviour of the world. What science says is possible or impossible is based upon precedent. That doesn't mean to say that something that surprises us – such as the virgin birth – can't happen.

*People often describe such surprises as 'miracles'. What, do you think, is a miracle?*

A miracle is very difficult to define but I'd say it is a wonderful event which is a sign from God. And yes, miracles do still happen today.

*Increasingly people today are looking to science to provide 'miracle' cures. Are you worried that scientists – through the likes of fertility treatment, gene therapy and the implanting of human DNA into animal cell shells – could end up playing God?*

The term 'playing God' is used in rather strange ways. In science God has given us certain abilities that are to be exercised with moral caution. There are things, such as Nazi medical experiments, that one human should not do to another. There are debates about when life itself begins. These things must be handled with care. The Christian perspective is: God has made all these things, so what should we be doing with them? And that's where the ethical and moral debates start. Good motive is one thing – everybody wants cures, everybody wants to alleviate suffering. But what we do should be guided by the fact that we are responsible to God for how we use what he has made. This is his world and we need to be careful. I would hate Christianity to be seen as a negative influence but scientists – and society at large – shouldn't rush into things just because they are do-able. We need to exercise caution based on our responsibility towards God.

*In* The God Delusion *Richard Dawkins[1] writes: 'The existence of God is a scientific hypothesis like any other.' Can scientific method prove the hypothesis that God exists?*

Science can deal only with matter and energy. Clearly, we can't take a piece of scientific apparatus to determine

God. Evidence for God, as I said earlier, is of a different kind. The existence of God is not amenable to science. Time and space are part of the natural world. God is 'beyond' time and space. If you're looking for mystery in science, consider this: there was no time or space before the Big Bang. The very phrase 'before the Big Bang' is a contradiction. The Big Bang brought time and space into being. There wasn't a 'before' for some other 'being' to create God. In fact, if you concluded that God had a creator, that creator would be God by definition.

In the fourth century Augustine said that creation was with time, not in time. He argued that if God created everything, he must have created time. The position today is founded on the much firmer scientific footing that time is part of the creation. But the proposition is mind-boggling. That's one of the reasons people get gripped by science. They want to understand more about the beginnings.

*Speaking of beginnings, what does science say about creation?*

Creation is a theological concept. Science studies the origins of matter and energy. They are not the same thing. Theologically, 'creation' is God's act of bringing-into-being everything that there is and sustaining it in being. Science studies the processes that resulted in the universe. Science looks at how the universe developed, but such a question as 'Is the universe designed or accidental?' is not a scientific question. Rather it is a question, in the words of 'Science in the National Curriculum' guidelines, 'that science cannot address'. Someone saying: 'I used to believe God created the universe but now I know it was a Big Bang' is a nonsense. God explains the 'Who?'; Big Bang explains the process – the possible 'How?'

*The media is increasingly running stories about creationism, especially in relation to the teaching of creationism in schools. What is creationism?*

No one can believe that God created the universe and everything in it without being what might be called a 'creationist'. But since about 1980 the word 'creationism' has come to be associated in the public mind with something which is properly termed 'young Earth creationism'. Broadly, young Earth creationism is characterised by two beliefs. The first is that God created the heavens and the Earth. This is something all Christians, Muslims and Jews believe. The second is that God did it in a short time in the geologically recent past – somewhere in the region of ten thousand years ago.

The statement 'God created' is not a scientific concept. When and how the universe might have come into being is. Given the findings of cosmology, physics, biology and palaeontology, I'm convinced that the Earth is around 4.6 thousand million years old and the universe about 13.7 thousand million years. I do, however, firmly believe in creation – that it was an act of God that brought everything into being and that it is God who sustains it, moment by moment. If one accepts, in Francis Bacon's terms, that God is the Author of the Bible and of Nature, then one would expect there to be harmony between them when they are properly understood. This means we have to read and interpret both books with care.

*How, then, should we read the creation accounts in Genesis 1 and 2?*

If we read 'day' as meaning a 24-hour period measured by the rise and setting of the Sun due to the rotation of the Earth, we have a problem. In Genesis 1 the Sun, Moon and stars aren't created until the fourth day. So what does the word 'day' mean on the first three days? Early Church Fathers such as Origen and Augustine realised that the word 'day' was being used differently. Origen wrote in about AD 225: 'What man of intelligence, I ask, will consider as a reasonable statement that the first and the second and the third day, in which there are said to be both morning and evening, existed without Sun and Moon and stars, while the first day was even without a Heaven?' He and others understood that 24-hour days with mornings and evenings were meaningless until the Sun and other stars were created on day four.

Augustine, who lived from AD 354 to 430, was more outspoken still. In his book *The Literal Meaning of Genesis* he wrote

> Usually, even a non-Christian knows something about the Earth, the heavens and the other elements of this world … Now, it is a disgraceful and dangerous thing for an infidel to hear a Christian, presumably giving the meaning of Holy Scripture, talking nonsense on these topics … People outside of the household of the faith think our sacred writers held such opinions, and … if they find a Christian mistaken in a field which they themselves know well and hear him maintaining his foolish opinions about our books, how are they going to believe those books in matters concerning the resurrection of the dead, the hope of eternal life, and the Kingdom of Heaven?[2]

There are many different writing styles in the Bible, including metaphor and allegory. Sometimes there is a mixture of factual content and symbolism. We don't know all the answers. Some things we need to put in a pending tray until we get the answer, but for me Genesis 1 speaks primarily about God who is the Creator.

*Let's look at Adam and Eve. Did they really exist or are they allegorical figures?*

Genesis uses 'Adam' as a person's name and as a generic term for humankind. The question is: When is

'Adam' used historically and when symbolically? For instance, is the story of the creation of Eve from Adam's rib while he is asleep an historical event or a symbolic account? It has been said that Eve was not taken from Adam's head to lord it over him, nor from his feet to be trodden underfoot, but from his side to be equal with him and from under his arm to be protected by him. These early chapters have symbolism in them. The tree of the knowledge of good and evil, whose fruit God forbade Adam and Eve to eat, is not a biological tree. The account is an indication of humankind's rebellious attitude towards God and its consequences, and that God's best plan is for humankind to live in close harmony with him. The overriding truth of the Bible is that God created the universe. It did not come about by accident.

*Do you think that belief in God as Creator rules out believing in the theory of evolution?*

No. Some evolutionists, though, wish to try to rule out the idea of creation. Charles Kingsley liked Darwin's ideas. He said people would now have to choose between a master-magician God and one who is a living, ever-working God. Darwin called evolution 'descent with modification'. This is universally accepted. In general, children look like their parents but they have personal modifications. Without evolution, there would be no diversity.

*Is it wrong to say, then, that the universe came about by evolution?*

We need to be careful. There are different types of evolution – stellar evolution, chemical evolution, organic evolution and so on. But we can't say that evolution brought the universe into being. That's like saying automation brought a car into being. Automation is the process. A design team creates a car. Evolution is a process, not an act. It is the 'how', not the 'who'.

Evolution doesn't *do* anything; that is God's action. Evolution is a term to describe the process taking place – that is 'descent with modification'.

*Historically, why have some people thought – why do some Christians still believe – that evolution conflicts with God as Creator?*

Part of it may be because of what it did to the argument that William Paley put forward. He saw the marvellous fit between something and the job that it did (a hawk catching a bird, for example) as evidence of God creating each thing for the purpose that it had. When Darwin's *The Origin of Species* was published, those who were unsympathetic towards religion saw natural selection as an explanation which displaced a belief in divine design. It did nothing of the sort, however. It simply indicated how God achieved adaptation in the process of descent with modification.

I don't see any reason why we should be worried about our ancestry. We are part of the animal kingdom. Evolution does not teach we came from apes but that apes and humans have a common ancestor. There's no reason to be unhappy about that. After all, if, as the Bible says, 'dust you are and to dust you will return', it really doesn't seem to matter.

Science looks at physical similarities and differences between apes and humans. The Bible says humankind is made in God's image. To me, that speaks about an ability to relate and respond to God. It is about our spiritual nature rather than our physical form. It says that God considers us precious.

*How compatible is the Big Bang theory with believing the biblical account of creation?*

'Big Bang' is scientific shorthand for a 'singularity' in which the ordinary laws of physics are unlikely to have held. Time and space came into being with the Big Bang.

This is what scientists talk of as 'origins'. Creation is that act of God which brought everything into being – by whatever means. To me it's important that God created the universe because it's about truth. The Bible teaches that God created everything, including humankind, whom he loves and for whom he has provided redemption from sin.

Genesis says that God made Adam out of dust. Science seems to have a similar view as it says we are made of dust from imploding stars. How can we be made of star dust? It is reckoned that after the first three minutes of the universe, matter consisted of hydrogen and helium. Gravity then drew these gases together. The closer they came, the higher the temperature rose. Eventually they burst into incandescence and stars were born. The temperatures were so huge and the energies were so high that the nuclei crashing against each other fused and built up into heavier elements, such as carbon, nitrogen and oxygen – the basic building materials of life. Some stars imploded at the end of their life, scattering these elements into space. The process then repeated on second-generation stars. Gradually, the atoms of life were built up. Our bodies are made up entirely of these atoms. If we want to sound romantic, our bodies are 'star dust', 'the ashes of long-dead stars' if we don't – or, to sound really grim, our bodies are 'reprocessed nuclear waste'. But there is not a nucleus of a carbon atom in our bodies that wasn't made in a star. That, to me, really is awe-inspiring.

## FACT FILE

*Michael Poole*
BSc (Physics), King's College London.

MPhil in Science and Religion in Education, King's College London.

Fellow of the Royal Society of Arts.

Associate, King's College London.

Head of Physics, Forest Hill Comprehensive School 1966–1970.

Overseas broadcaster on science and religion 1970–1973.

Lecturer in Science Education, King's College London 1973–1991.

Visiting Research Fellow in Science & Religion, King's College London 1991–present.

Author of some eighty articles and several books on the relationships between science and religion, including *User's Guide to Science and Belief* (Oxford: Lion Hudson, 2007).

# 5

# Dr John Lennox

## Mathematician

To the uninitiated, the formula to solve a quadratic equation or Newton's laws chalked on a blackboard are little more than mumbo-jumbo. To Oxford mathematician Dr John Lennox, however, mathematics is more than a thing of beauty – it is a sign that God exists. After studying mathematics at Cambridge, he became Reader in Pure Mathematics at Cardiff University. He is now Reader in Mathematics at Oxford University and Fellow in Mathematics and Philosophy of Science at Green College, Oxford. John is an expert in group theory ('an abstract algebraic method that helps us cope, among other things, with symmetry, and that is used – for example – in atomic physics'). He also has a degree in bioethics. He was born and raised in a Christian family in Armagh, Northern Ireland.

*John, how for you does maths point towards the existence of God?*

There is no formula that says x+y=z therefore God exists. It is more the fact that we can do mathematics and that it corresponds to something out there. Galileo said that mathematics is the language God has given us so that we can describe the world. Mathematics is about being able to compress observations of nature into a single formula, as Newton did with his law of gravity. Mathematics and science don't do away with God. When Newton discovered gravitation, he didn't say, 'Now that I know how it works I don't need God.' In fact he said the exact opposite: 'What a marvellous God who does it like this!' Newton wrote *Principia Mathematica* so that the sheer wonder of mathematical descriptions of the universe might provoke the thinking person into believing in God.

*How were you provoked into believing in God?*

My parents were evangelical Christians, and so as a child I was taught the Christian faith. Despite living in a sectarian society, my parents encouraged me to think about faith for myself. They believed one had to come to one's own conviction when one was mature enough, although I personally cannot recall a time when I didn't believe in God. In my first week as a student at Cambridge a student asked me if I believed in God. Then he said: 'Oh, sorry, you're Irish; you all believe in God and fight about it.' That was a turning point. I had to work out whether my faith was simply hereditary or whether it was true. As part of the process of finding out, I've been discussing with atheists ever since.

*You have faced Richard Dawkins in live debate on whether God is a delusion and addressed atheism in your book* God's Undertaker: Has Science Buried God?[3] *How do you view the science/faith debate?*

The key question is whether the physical universe is all that there is or whether there is a Creator. The answers will depend on which of three worldviews one holds: atheism, which is often expressed as naturalism or materialism; pantheism, which equates God with nature; and theism, which says there is a God. My question is: With which of these views does science sit most comfortably?

I suggest that if you take God out of the equation, science itself loses some of its *raison d'être*. Science isn't neutral. Science positively supports belief in God. The great sixteenth and seventeenth-century pioneers of science were all believers. It was belief in a Creator which gave science its initial impulse. The atheistic viewpoint is that the universe comprises nothing but matter and energy, and that everything, including life itself, is the product of mindless unguided processes. This means our thoughts are similarly the results of such processes. If that's the case, why should we believe the theories that our thoughts spin? Atheism therefore undermines the belief in the rational intelligibility of the universe without which no science can be done. Biblical theism, however, validates that intelligibility, in that it tells us that the reason why we can begin to understand the universe is that God created both it and the human mind.

Science has immense cultural authority. After all, it has been enormously successful in producing technology that we all enjoy. So people look to science to provide all the answers. Bertrand Russell once said that 'what science cannot discover, mankind cannot know'. That's logical nonsense. For, since it is clearly not a statement of science, we cannot know it is true, according to what it itself says. People are rightly passionate about science but it does science no service to suppose that science can deal with every aspect of reality. Nobel prizewinner Sir Peter Medawar points out that science has its limits. It answers questions about the universe and how it works but it can't answer these simple questions like: Where do we come from? Where are we going? What is life about? Nor can it deal with morality.

*Why then do people see faith in God and science as conflicting?*

A lot of that stems from a caricature of faith. Some scientists, including Dawkins, define religious faith as blind faith – belief without evidence. This conflicts with their perception of science as something based on evidence. I would argue that the New Testament helps us to understand very clearly that faith in the biblical is also a commitment based on evidence.

*Where, as a man of science and faith, do you find evidence for God?*

I can't prove God in the way that I can prove a mathematical theorem, but I can find evidence beyond reasonable doubt. Firstly, the created universe. The beauty, magnificence and complexity of the universe points not to mindless unguided processes, but to a Creator. Secondly, the existence of moral conscience points to an objective standard. In his letter to the Romans, Paul writes about the light with which God illuminates people. We all believe there are moral absolutes – that, for example, torturing babies is wrong. Where does conscience come from, if not from God?

Thirdly, DNA. Inside every one of the human body's ten trillion cells there is a digital database, encoded in some 3.1 billion letters of DNA. As a mathematician, I do not find it convincing to believe that the language-like structure of DNA has come about merely through random unguided processes. The very existence of this database is totally consistent with the biblical view that says: 'In the beginning was the Word . . . Through him all things were made; without him nothing was made that has been made' – the very opposite of materialism.

Fourthly, I am a Christian because I believe there is evidence that Jesus is God incarnate – that God has encoded himself in human life in the person of Jesus. The key evidence for that is that Jesus rose from the dead as a matter of history. The three facts of the empty tomb,

Christ's subsequent appearances and the meteoric rise of the Christian Church from a bunch of dispirited people whose leader had been crucified to a vigorous worldwide movement can together be explained only by the resurrection.

Fifthly, I'd point to the desire to know – whether that's expressed through science or faith – as evidence that humankind is made in the image of God.

*What does being made 'in the image of God' mean?*

To me, it means we have been made with the capacity to have a relationship with our Creator. God revealed himself to humankind as a person – Jesus Christ. God is creative; we are creative. God looks for relationship; we have the capacity for relationships.

*John, why is it important to you that God created the universe?*

Each of us wants to know our identity. A person without a past is a person without an identity. That's why people are fascinated by the big story science is offering. They want to know who they are. The idea that my personal origin is in the mind of a Creator, who is interested in me and has created me in his image, gives me colossal dignity and sense of value.

The psalmist says: 'The heavens declare the glory of God.' But the heavens weren't made in his image, only humans were. That attributes immense value to human beings which atheism cannot parallel. The dignity of human beings as made in the image of God seems to me to be one of the most liberating messages we could possibly hear.

*Some people see a Bible-based belief in a Creator-God and Darwin's theory of evolution as incompatible. Do you?*

Historically it suited those scientists who wanted a priesthood of science to have people believe that Darwin's theory has eliminated God. It hasn't, since it logically can't. In any case, micro evolution – variations between generations within a species – is not controversial. Today's generation, for example, is taller than my generation (mainly because of access to better diet). As a mathematician I can see evolution as an optimising process but it doesn't account for the existence of life and Darwin never claimed that it did. Evolution works only when you have got life to start with.

*Increasingly, science suggests that the universe – and subsequent life – came into being through the Big Bang. Is this view compatible with the Bible?*

'Big bang' is a label on a mystery. Physicists talk in terms of a 'singularity', where the laws of physics break down. The Bible, too, describes a singularity: 'In the beginning God created the heavens and the earth.' The very fact that science points towards a beginning to space-time confirms that Genesis has been right all along. It could be that there was more than one singularity: the creation of the universe being one, the origin of life being another. Again we see echoes of this in Genesis 1, where 'And God said' occurs a number of times. I do not believe that human life came about by purely unguided mindless processes but that inputs of information from outside the universe were involved – in other words, singularities occasioned by God.

Nor did the singularities stop at creation. The virgin conception of Jesus, his incarnation and resurrection were singularities with immense implications which are the foundations of the Christian faith. The big question is: Is the universe created or not?

## FACT FILE

*Dr John C. Lennox*

MA (Mathematics), Cambridge.

MA (Bioethics), University of Surrey.

PhD (Mathematics), Cambridge.

DPhil (Mathematics), Oxford.

DSc (Mathematics), University of Wales.

Reader in Pure Mathematics in the University of Wales 1985–1997.

Reader in Mathematics in the University of Oxford 2006–present.

Fellow in Mathematics and the Philosophy of Science, and Pastoral Advisor at Green College, Oxford.

Lecturer in Science and Religion at Wycliffe Hall, Oxford.

Senior Fellow of the Trinity Forum.

Author of more than seventy mathematical papers and of books including, *Informetika* (Budapest: Harmat, 2001), *The Definition of Christianity* (with David Gooding, Port Colborne, Ontario: Gospel Folio Press, 2004) and *God's Undertaker: Has Science Buried God?* (Oxford: Lion Hudson, 2007).

Co-author of two research level texts in algebra in the Oxford Mathematical Monographs series: *Subnormal Subgroups of Groups* and *The Theory of Infinite Soluble Groups*.

# 6

# Dr Francis S. Collins

## Gene pioneer

As a geneticist, he has peered down a microscope and been the first person to see processes that previously only God knew about. As a physician, he has been the last person to care for a dying patient. As a Christian, Dr Francis Collins has a perspective on life and death.

In the early 1980s, he became the first person to discover a single letter of the human genetic code that modulates production of foetal haemoglobin. In 1989, he led the team that discovered the genetic misspelling that causes cystic fibrosis. Four years later, he became director of the National Human Genome Research Institute in America and headed an international race against time and commercial interests to sequence the 3.1 billion letters of the human genome.

On 26 June 2000, he stood beside Bill Clinton in the White House as the President pronounced the first survey of the human genome ninety per cent complete. 'Today,' said Clinton, 'we are learning the language in which God created life.' It was a phrase borrowed from Collins.

*Francis, what is the human genome?*

The human genome is the entire collection of the DNA of our species. Human DNA carries information by a series of chemical bases. This information has only four letters in its alphabet – A, C, G and T.

In the double helix of DNA these letters appear like rungs on a ladder in four possible pairs: A-T, T-A, C-G and G-C. Because they combine only in this way, if you split all the pairs in half, cutting the ladder down the centre of each rung, each half-ladder contains all the information needed to rebuild a complete copy of the original.

DNA is a bit like a software program that sits in the nucleus of a cell. A particular instruction – a gene – comprises hundreds of thousands of letters of ACGT code.

The whole of the human genome is 3.1 billion of those letters. Such is the complexity of the information carried within each of the one hundred trillion cells of the human body, that if someone read the code out loud at three letters per second, it would take thirty-one years to complete.

The human genome is humankind's instruction book – the hereditary code of life. For a believer like me, it is the language of God.

*What was the Human Genome Project?*

The main goal of the Human Genome Project, which started in 1990, was to map those 3.1 billion letters. The project was completed (ahead of time and under budget, I'm proud to say) in April 2003.

But that was only the foundation. Just looking at the long string of As, Cs, Gs and Ts that direct our cells doesn't mean we know how it all works. We are now engaged in the most exciting part, figuring out how and where misspellings can contribute to ill-health or disease, and how to bring about cures.

We are carrying out genetic investigations into common conditions such as diabetes, asthma, high blood pressure, mental illness, Alzheimer's disease and

Parkinson's disease. In the relatively near future we may be able to tell people their individual risks of contracting or developing these illnesses, and give them a chance of reducing those risks.

*You discovered the gene in which cystic fibrosis (CF) occurs. Have you devised a gene therapy to cure the disease?*

My team and I, collaborating with another group in Toronto, discovered the CF gene in 1989. The search took seven or eight years. It was a tedious process – like looking for the proverbial needle in a haystack, except at the beginning we didn't even know for sure which haystack to look in.

In 1985 it was demonstrated that the CF gene must live somewhere within a two million base-pair segment of DNA on chromosome 7. We were in uncharted territory. Then one night in May 1989 my Toronto colleague and I read the day's data from the lab – a deletion of just three letters (CTT) from the DNA code in the protein-coding part of a previously unknown gene had to be the cause of CF.

We are hopeful that one day CF will be a thing of the past. But people underestimate the time it takes to bring about cures. There are a lot of steps that have to follow that kind of initial discovery.

You have to understand what the normal function of this gene is, what is wrong with it in people with CF. Then you need to figure out how you can compensate for that, either with a gene therapy or a drug therapy. Only now, eighteen years later, are we seeing new drugs coming into clinical trials that are based specifically on the knowledge of how the CF gene works.

*So not all genetic diseases are cured by engineering a patient's DNA under a microscope?*

All disease has some hereditary contribution. Some diseases, such as CF, are due almost entirely to genetic misspellings. Other diseases, such as diabetes, come about as a combination of some misspellings that place somebody at risk, plus environmental circumstances such as diet and exercise.

Finding the genetic basis of a disease leads to two possible treatment pathways – to use the gene itself as the treatment (gene therapy) or drugs. For example, haemophilia is caused by a defective blood-clotting gene. There has been some success in introducing a normal copy of that gene into a patient's liver and getting the blood starting to clot again. But most gene therapy efforts are not yet successful.

The problem with CF was to figure out how to deliver normal copies of the gene into a patient's airway without triggering the immune system which would try to repel it as a foreign body. That turned out to be extremely challenging. So we are coming around to the perspective that the solution may be to use what we know about a gene and then design a drug that is very precisely tuned to the problem.

*What can gene research do for conditions such as Down's syndrome?*

Somebody with Down's has an extra copy of chromosome 21, so has forty-seven chromosomes in total, instead of the usual forty-six. This means that all of the genes on that chromosome are present in three copies instead of the usual two. The idea that a doctor could remove the extra chromosome is difficult to imagine at present, as every one of the one hundred trillion cells in that person's body has that extra chromosome.

Maybe the strategy is to focus on the effects on the brain. Somewhere in the collection of genes on chromosome 21 must be some that are very sensitive to 'dosage'.

If we knew which genes were involved, we might be able to devise a drug that would dampen down their production, or compensate for the overdose they create.

Treatment would have to start in the early days of brain development. So there may come a time when we could make the consequences of Down's milder or even reverse them. But that's a long way off.

*What role do genes play in governing our behaviour? Is there a selfish gene or a gay gene?*

There are undoubtedly genetic variations that play some role in behaviour, as the study of identical twins (who have the same DNA) has indicated for a long time. A few clues about behaviour, such as anxiety, have already been discovered. We have looked to see whether genes have a part in determining sexual orientation, but so far no specific genes have been identified. The indication is that sexual orientation is genetically influenced but not hard-wired by DNA.

Behaviour is not the major focus of genome research. The big push is for medical benefits. We are concentrating on finding the genetic factors in diabetes, cancer, heart disease and mental illness.

*Conditions such as cystic fibrosis are caused by a letter in the genetic code being out of place. How can an all-loving Creator let that spelling mistake happen and allow a person and their family to suffer as a consequence?*

That is a tough question. I don't think God makes the spelling mistake. I think God's heart is also broken by the occurrence of these diseases. (The fact that God sent his only Son to suffer for humankind means that Christians are perhaps in a special circumstance of knowing they have a God who appreciates what suffering is.) Of course, much human suffering comes about because of what we humans do to each other. But what about a terrible earthquake or cancer in a child?

The English physicist Professor Sir John Polkinghorne asks the question: Why does God allow this kind of suffering? Does it mean God isn't loving enough or that he isn't powerful enough? Polkinghorne's answer is that God in his wisdom chose to create the universe, the world and ourselves, using a process over very long periods of time that involve change – for example, tectonic plates moving which cause earthquakes. God also chose to use a mechanism that requires DNA to change over time, without which there would be no life. This is the same mechanism that, on rare occasions, results in a mis-copying of a critical gene and produces a disease like cystic fibrosis or cancer. Polkinghorne argues that you can't have one without the other.

Ultimately, I think we have to come to grips with the fact that all suffering may have a significance beyond what we can appreciate. I know that the times I have learnt most about myself and my relationship to God are not those where everything was going great but the times when there have been serious difficulties and suffering.

C.S. Lewis said suffering is God's megaphone to rouse a deaf world. Perhaps then we shouldn't expect that the kind of world God has in mind for us is one where at the end of every day it could be said that everyone had a good time.

*Talk of cloning and genetic modification sounds exciting because we want to be rid of life-threatening conditions. But it also sounds frightening – scientists playing God, designer babies, society defining (as in the film* Gattaca*) who is valid and who is invalid on the basis of genetic perfection. What ethical safeguards are there to the work on the human genome?*

The Human Genome Project distinguished itself from the beginning by committing a significant proportion of its budget to studying the ethical, legal and social issues. That scholarship led to the conclusion that a major issue in need of resolution is protection against genetic discrimination: you didn't get to pick your DNA sequence, and so it shouldn't be used against you in determining healthcare or your qualifications for a job.

After twelve years of many of us trying to raise consciousness about the need for this kind of protection, this year the US Congress is very close to passing a Bill that would prevent genetic discrimination. For people who could be denied employment or medical insurance if details of their genetic make-up were made public, this is a critical issue.

Scientists have something to contribute to those debates because they know the facts of what is possible and what isn't. But they also have a natural conflict of interest. They want to see science going forward, so the decisions about what boundaries should be put around research should not be made by scientists alone. There has to be broader public discussion. What makes me a little crazy are claims by some people that the whole area of genetic research is dangerous and should be stopped. This is the most unethical stance imaginable.

If we, as a benevolent society, are called to try to alleviate suffering and to do something about the child with cystic fibrosis or cancer, then it seems to me it would be profoundly unethical to say: 'We haven't worked out some of these social consequences yet, so we're going to put genetic research on hold.' Try saying that to a parent who has a dying child!

*Is science, much of which seems to assume that there isn't a God, irreconcilable with faith?*

Some people see science and spirituality as opposing forces – somehow mutually inconsistent. They are not. I find them entirely complementary. I don't think science will ever be in a position to prove the existence of God. But I do think there are evidences that come out of science that point more towards than away from the existence of a Creator God who cares about human beings.

Science has limits. There are huge questions science cannot answer, such as 'Why did the universe come into being?', 'Why are we here?' and 'What happens after we die?'

*When you look down a microscope at the building blocks of life, what evidence for God do you see?*

The fact, for instance, that the universe had a beginning (and Big Bang theory calls out for an explanation, because nature has not been observed to create itself) immediately draws the question of how that beginning came to be. At the time of this incredible explosion of matter and energy there must have been something outside nature that was capable of creating. And that 'something' sounds like God himself.

When you look at the fine-tuning of the universe – the so-called anthropic principle – where you see how incredibly improbable it is that all of the constants that govern the behaviour of matter and energy were set at exactly the point at which some sort of stable life was possible, you can't help but marvel and conclude that there was something more than happenstance behind the process to set those constants in just that precise way.

Then there's the moral law, or 'law of right behaviour', which is present in humankind. We might argue about what is right and what is wrong but we all have a sense of right and wrong.

The moral law cannot be readily explained in terms of evolution because it sometimes calls us to do things that are really quite the opposite of what evolution would ask – like jump into a river to save a drowning stranger. The moral law has been fully set in the heart of only one species – humankind. There's no bigger signpost towards the existence of a personal God.

*How did you become a Christian?*

I wasn't raised in a faith tradition. As a child I was vaguely aware of the concept of God. My parents enrolled me in a church choir to learn music. I took to the music but not to the theology.

As a student, both at college and as a physical chemistry graduate student at Yale, it was convenient for me to walk away from any responsibilities other than the ones I chose for myself. I was surrounded by people in science who largely ignored any spiritual aspect of life, so I just ran with the crowd.

I was convinced that everything in the universe could be explained by equations and physical principles. No thinking scientist, I concluded, could seriously entertain the possibility of God without committing intellectual suicide. I had gone from being an agnostic to an atheist. To me, faith was sentimental superstition.

But then I changed direction and went to medical school, and I started to encounter the reality of suffering and death in a much more up-close-and-personal way.

One afternoon an elderly patient with a terminal illness shared her faith with me and explained why this gave her comfort and peace as she saw the end of her life approaching. She turned to me and said: 'Doctor, I've told you about my faith but you haven't said anything. What do you believe?'

I had never been asked that question so directly. Suddenly I felt a great sense of unease that I didn't have a good answer, that I had never seriously asked myself the question, that as a scientist I had never taken the time to look at the evidence for and against belief, and had drawn a conclusion without having considered that evidence.

I started by visiting a Methodist minister. He listened to my confused questions and gave me a book to read. It was *Mere Christianity* by C.S. Lewis. As I read it I realised that all my arguments against the rationality of faith were those of a schoolboy. Lewis's case for faith, which is something I had never encountered before, was extremely compelling.

I fought it. I didn't want it to be true. I was reading the book to shore up my atheism, not to become a believer. But, as Lewis puts it, the 'Hound of Heaven' was after me. I fought the evidence for almost two years.

*When did the searching stop and the finding begin?*

It became clear that I had to make a decision not only about whether to believe in God but also about what kind of God he was. I encountered the person of Jesus Christ as this remarkable figure in history who clearly was different from any other figure in any other faith.

Jesus claimed not only to *know* God but also to *be* God and to forgive sins, and he died on a cross in a way that took me a long time to understand. Eventually it made the most beautiful, perfect spiritual sense. I realised that this was not just a story to walk away from. I had to reach my own verdict based on the evidence.

My moment of commitment came one autumn day while I was hiking in the Cascade Mountains. It was a beautiful afternoon and as I rounded a corner I unexpectedly saw a frozen waterfall, hundreds of feet high. The remarkable beauty of creation was overwhelming. I could no longer resist. I fell on my knees and asked Christ to be my Saviour. My days of wilful blindness towards God were gone. The search was over.

*Do you believe in the virgin birth, the resurrection and the second coming of Jesus?*

I do.

*Even though these things are outside scientific rationalism?*

Yes. The main questions one has to ask oneself are whether you believe in a God who is outside nature, and whether you believe in Christ as divine. If you are willing to say yes to both of those questions – and I became unable to say anything but yes after considering all the pros and cons and feeling the spiritual tug in my heart – then the idea of these miracles is not really such a problem, even for a scientist. If God is outside nature and is all-powerful, there is no reason why he cannot occasionally stage an invasion of nature, which we would call a miracle, to get our attention.

*There are scientists who conclude that some theories have a negative consequence for religion. Richard Dawkins, for instance, claims that evolution is necessarily atheistic. Is that the way you see it?*

No. There are a number of areas where people are being asked to make a choice between either science or faith. Often the science involved is evolution. The atheistic approach, which Dawkins takes, claims that science trumps faith. To draw the conclusion that an acceptance of evolution requires an acceptance of atheism as a personal theology, as he does, is simply to go outside the evidence.

Evolution is an incredibly unifying approach to understanding the relationship of organisms. Study of DNA strongly supports Darwin's proposal of descent from a common ancestor. By comparing the genomes of many different organisms, we can see how this fits precisely with what one would expect if they had begun as a common ancestor and then diverged over very long periods of time.

Evolution says only how the process might have occurred. It doesn't say a thing about why life is here. It doesn't say much about whether our existence on this planet has a purpose beyond randomness. For those answers, we must look beyond science.

The creationist view is that faith trumps science. It argues that evolution is a lie. It says that the relatedness of organisms as visualised by the study of DNA is simply a consequence of God having used some of the same ideas in his multiple acts of special creation.

Confronted with such facts as the similar ordering of genes across chromosomes between different mammalian species, or the existence of repetitive 'junk' DNA in shared locations along the DNA of humans and mice, they simply dismiss this as part of God's plan.

*Do you have a preferred personal view of the origins of the universe?*

I do not see it as a choice between faith in a personal God and science. Many materialists, noting the advances of science filling in the gaps of our understanding, announce that God is dead. Many believers see the advance of science as dangerous and untrustworthy – a threat to God. Both positions are dangerous and unnecessary. The God of the Bible is also the God of the human genome. He can be worshipped in the cathedral and in the laboratory.

God's creation is majestic, awesome, intricate and beautiful. Creation cannot be at war with itself. Only imperfect humans start such battles. Only we can end them.

For me, they can be blended without compromise through what is known as theistic evolution – what I prefer to call 'Bios through Logos' (Life through the Word), or just 'BioLogos'.

Basically, this says that God, who is unlimited by time or space, created the universe and established the natural laws which govern it. He then chose evolution as the mechanism to create microbes, plants and animals.

God then intentionally chose the same mechanism to give rise to special creatures who would have intelligence, a knowledge of right and wrong, free will and a desire to seek him – humankind. He prepared the biological home for humanity by evolution, but then he filled it with the soul, the knowledge of right and wrong (the moral law) and free will.

God also knew that these creatures would ultimately choose to disobey the moral law, and thus the fall would occur. His provision for this would be to send his Son, Jesus, to live, die and be raised for our salvation.

## FACT FILE

*Dr Francis S. Collins*

BS (Chemistry), University of Virginia 1970.

PhD in Physical Chemistry, Yale 1974.

MD, University of North Carolina 1977.

Residency in Internal Medicine, North Carolina Memorial Hospital 1977–1981.

Fellow in Human Genetics, Yale 1981–1984.

Professor of Internal Medicine and Human Genetics, and Investigator, Howard Hughes Medical Institute, University of Michigan 1984–1993.

Director of National Human Genome Research Institute, National Institutes of Health 1993–present.

He led an international team that accomplished the sequencing of the entire human genome in draft form in June 2000, and in final complete form in April 2003.

He now leads U.S. research to apply genomic discoveries to advances in medicine.

Team leader of discoveries of a number of genes and genetic risk factors, including those responsible for cystic fibrosis, neurofibromatosis, Huntington's disease, Hutchinson-Gilford progeria syndrome (a dramatic form of premature ageing), and adult onset diabetes.

Author of *The Language of God: A Scientist Presents Evidence for Belief* (New York: Simon & Schuster, 2006).

# 7

# Professor Norman C. Nevin, OBE

## Medical geneticist

Abortion, cloning, disease, suffering. Not the Four Horsemen of the Apocalypse but certainly among the major challenges of the age. They cut to the quick of our humanity, its wonder and its fragility. Arguments about their rights and wrongs are just as likely to generate heat in Parliament and church pews as in doctors' surgeries and pub lounges.

Few people have thought longer and harder about the rights and wrongs of medical science than Norman Nevin OBE, Professor Emeritus of Medical Genetics at Queen's University, Belfast. For twenty-five years he was Professor of Medical Genetics at Queen's University and from 1993 to 2006 was member and chair of the United Kingdom Gene Therapy Advisory Committee. When Norman entered medicine in 1953 – the year Watson and Crick discovered the structure of DNA – medical genetics was very much in its infancy. Over the years, his research interests have been investigating the causes of congenital abnormalities, such as spina bifida, the commonest congenital abnormality affecting new-born infants. He was involved in the identification of the importance of vitamins and folic acid before conception in the prevention of spina bifida and related abnormalities. His other major area of research is the molecular basis of single gene disorders, affecting muscle, bone, skin and the eye. For more than twenty-five years he was vice-chair of the Research Ethics Committee of the Faculty of Medicine at Queen's. In 1970 he was a founder member of the UK Clinical Genetics Society and served as its president from 1991 to 1992. In 2003 he was appointed OBE for his contribution to gene therapy research.

*Norman, where did your passion for medicine come from?*

My mother was a nurse and as a boy of eleven or twelve I wanted to be a doctor. I took a year out to do an intercalated honours BSc in anatomy and qualified with honours in medicine in 1960. After a year as a houseman in Belfast's Royal Victoria Hospital, I undertook a doctorate in pathology. It was when carrying out post-mortems on babies with congenital abnormalities that I felt I should branch out into the newly developing field of clinical genetics. I then went to London's Institute of Child Health and to the MRC's Population Genetics Unit at Oxford before returning to Belfast, as lecturer in human genetics at Queen's.

In 1967 I set up the genetic counselling service in Northern Ireland. This was a clinical service where couples who had a child with an abnormality or a genetic disease could talk the issue through with a specialist and receive advice about the management of the child, and about possible risks to future children. There were no cures to offer because at that time little was understood about these conditions. It was only in 1959 that Down's syndrome was identified as a chromosome disorder. I also established ground-breaking laboratories where chromosome and gene disorders in babies could be diagnosed. The introduction of new technologies – scans and amniocentesis – resulted in new ethical challenge. How would a couple respond when it was discovered that the mother was carrying a baby with an abnormality or a genetic disease? Such a situation raised the dilemma of abortion. It still does.

*And how, as a Christian physician, do you view abortion?*

As a doctor, I don't see abortion as a totally black and white issue. Sometimes it is, in a sense, the lesser of two evils. For example, when a woman develops a rapidly spreading cancer during pregnancy, the doctor is faced with the dilemma: Does he treat this woman, knowing that the treatment may severely damage the baby? Where a mother's life is at risk, and with the parents' consent, then the correct course of action would be to end that pregnancy. At the other end of the spectrum, the majority of abortions in the UK today are carried out for so-called 'social' reasons. I am not in favour of abortion on these grounds. It must be emphasised that only one to two per cent of abortions are carried out because of a gross abnormality in the baby.

As a Christian, my approach is that when the couple have been given a result which indicates that the unborn baby has a serious physical or mental handicap, they must have careful counselling which should include the option of terminating the pregnancy. The decision to terminate must be that of the couple. I respect those individuals who under no circumstances would have an abortion. However, I would be wrong if I did not apprise them of the option of termination. My role as a genetic counsellor is to provide couples with the most up-to-date and accurate information to enable them to make an informed decision. Whatever decision couples make, they need care and support. I am called upon to show compassion and to have empathy with those who are suffering; that's how I square my Christian faith with this dilemma.

The 1967 Abortion Act does not apply to Northern Ireland but, under case law, women can have terminations in the Province if, in the opinion of a clinician, the foetus is seriously abnormal. There are conditions – like a foetus with no brain (anencephaly) – where it is easy to make that decision to terminate the pregnancy. I have no problem with that. Technology is now so advanced that it is possible to detect many disorders which are less serious or can be adequately treated after birth. This raises the problem of where to draw the line. The prenatal diagnosis of Down's syndrome, for example, presents a major dilemma for would-be parents. Should they continue, knowing their child may be helped towards independent life, or should they have a termination? Which conditions to terminate can never be laid down in legislation. The questions may only be resolved by advising the couple of the likely long-term development and possible future treatment of the child.

*When you started in practice, doctors did not talk about cures for genetic diseases. Today the talk is about gene therapy. For some people gene therapy smacks of scientists playing God. How do you see it?*

In the Bible there is no complacency in the face of disease and suffering. God has given knowledge and wisdom to doctors and scientists to develop treatments for disease. Genetic disorders are no different from other diseases. As for 'playing God', I believe God expects doctors and scientists to use all this knowledge for the benefit of humankind. I don't think medicine will completely eradicate all abnormalities. The vast majority of abnormalities and genetic diseases occur without any family history. For example, one of the commonest causes of children who are slow in developing is that the mother drank alcohol during pregnancy. You are not going to eliminate that!

There are more than four-and-a-half thousand diseases which are due to a fault in the genes. The commonest gene disorder in the UK is cystic fibrosis, affecting one in every two thousand babies born. It is due to a faulty gene passed from each parent to the child. One in every twenty people in the UK carries that single gene. In the case of cystic fibrosis, a blood test enables doctors to identify whether a couple are carrying the defective gene. It also helps doctors make a more accurate diagnosis and, in time, hopefully, there will be a way to repair or replace defective genes.

In 2002 a boy with severe combined immune deficiency (SCID) was cured by means of gene therapy. Doctors took a sample of his bone marrow cells and inserted into these the normal SCID gene for which he was defective. The cells were grown for a few days in the laboratory and then re-infused into his blood system. Before treatment, the boy was prone to infections and had to be confined to a sterile room. His life span was limited to a few years. Now he is a normal young boy, able to go out and play with other children without the fear of 'catching an infection'. Haemophilia, too, is another disease where treatment with gene therapy is being developed. My dream is to see many more genetic diseases become amenable to treatment in this way.

*Norman, how would you describe your journey to faith?*

As I lad I regularly went to church, Sunday school and Boys' Brigade. (Today I'm president of the Northern Ireland Boys' Brigade.) Even though I was young, I realised I needed to make a personal commitment to Jesus Christ. I knew he was an historical figure. But I had to acknowledge that he died specifically for me and at the age of twelve I accepted Christ as my Saviour.

Being a Christian doesn't mean committing intellectual suicide. My faith helps me in my clinical and research work. The Bible doesn't provide chapter-and-verse answers to areas like abortion, prenatal diagnosis, therapeutic cloning or genetic engineering but it does gives some clear and guiding principles such as to 'Test everything. Hold on to the good' (1 Thes. 5:21) and to 'act justly and to love mercy and to walk humbly with your God' (Micah 6:8). I believe in a God of love. My faith gives me a personal, living relationship with Christ. When I face decisions or difficulties I can look to a higher authority for strength and guidance.

*Have there been particular occasions when you've needed divine strength and guidance?*

Suffering, pain, disease, illness are all consequences of man's initial disobedience to God (the fall) – part of man's nature. Seventeen years ago I had coronary artery bypass surgery. Ten days after I retired, I had surgery for bowel cancer. My immediate reaction was: Why me, God? I realise that just because I'm a Christian it doesn't that mean I'm immune to the problems of disease and suffering. It was a very difficult period in my life but God was with me throughout it. With the support of Jean, my wife, my family and many praying friends, I have come through that dark period of my life.

*Some people see the hand of God when looking into outer space through a telescope. What gets you awestruck?*

Every time I look down a microscope and realise the marvellous way in which man has been created, I feel drawn closer to God. The heart of a human cell is the nucleus, which contains most of one's genetic material. That genetic material is arranged on forty-six thread-like bodies – chromosomes. If the chromosomes are dissected out and laid end to end they would measure two metres. Each individual has about seventy quadrillion cells. If it was possible to lay all the chromosomes in the human body end to end they would stretch from the Earth to the Moon and back about eight thousand times. One individual!

Along the length of each chromosome are the genes – some twenty-five thousand. The genes are lengths of DNA made up of only four chemicals or bases. There are some three billion base pairs in the human genome. The DNA carries information about a person's hair colour, height, looks, etc, arranged in a unique code. If that information was printed in book form, it would occupy at least four thousand books each with five hundred pages. Is it any wonder the psalmist wrote that humankind is 'fearfully and wonderfully made'?

*Britain is one of the few countries to allow scientists to undertake research for possible genetic cures using human embryos. Throughout your career you've been involved with the ethical implications of genetic research. Do you approve of the use of human embryos for research?*

I believe it's a step too far. Although human embryos are a rich source of stem cells, they are not essential to obtain stem cells. Stem cells which can develop into nerve, heart cells or other cells can be harvested from adult human tissue such as bone marrow, teeth and fat. Research should focus on adult stem cell research and particularly on techniques of somatic (body) cell reprogramming, that is to induce differentiated adult cells (skin or bone cells) to revert to an embryonic cell state.

*And what are your thoughts on the ethics of cloning?*

Cloning is just a technique in which the nucleus of an egg is removed and replaced with the nucleus of a somatic (body) cell. This was the technique used by Professor Sir Ian Wilmut to create Dolly the Sheep in 1997. The question is: When is it permissible? From a moral point of view, I believe that attempts to clone a human ('reproductive cloning') should not be permitted. However, using cloning ('therapeutic cloning') to develop, for example, nerve tissues to treat conditions such as heart attacks, Parkinson's or Huntington's disease should be permissible with stem cells derived from adult tissue. In 2007, a team of researchers led by Professor Shinya Yamanaka of Japan succeeded in reprogramming mice skin cells to behave as stem cells. Indeed, Professor Wilmut is abandoning the cloning of human embryos as a source of stem cells and adopting the approach of the Japanese researchers. Scientists will always be faced with new boundaries. Today's and tomorrow's discoveries and their applications undoubtedly will raise new moral and ethical issues.

*Like designer babies? If doctors have the technology to identify and alter individual genes, is it only a matter of time before parents select personality and physical traits as well as asking for diseased genes to be eradicated?*

If we arrive at the stage when couples can select a baby for personality traits such as musical ability, higher IQ or 'good looks', scientists will have gone too far down the slippery slope. I believe that the research in genetics should be aimed only at ways of alleviating and curing disease, and at abnormality.

*In September 2007 the Human Fertilisation and Embryology Authority gave permission for scientists to combine genetic material of different species, such as using animal eggs from which the nucleus has been removed and human genetic material for the purpose of producing stem cells. Is this an acceptable alternative to the use of human embryos to produce stem cells? How concerned are you about the ethics of human-animal mixtures?*

As a Christian, I believe God created humankind in his own image and likeness (Gen. 1:26–27). This, to me, implies that humankind is distinct and unique from other animals. Only human beings have a special relationship with God. The fusion of human-animal genetic material would be counter to the sacredness of human life. Christians are not anti-science or anti-research. Indeed, I have been involved in research for most of my professional career. It is important to conduct research in an ethical and moral manner. The application of genetic means to treat and hopefully cure many diseases has enormous potential. I would sound a note of caution, though. In going down the road of genetic modification and manipulation of humans, we are in danger of ceasing to regard humankind as special.

## FACT FILE

*Professor Norman C. Nevin, OBE*
BSc (Anatomy), Queen's University Belfast.
MB, BCh, BAO (Medicine), QUB.
MD (Pathology), QUB.
House physician and surgeon, Royal Victoria Hospital, Belfast 1960–1961.
Lecturer/Consultant in Human Genetics, QUB 1961–1964.
Registrar in Medicine, RVH 1964–1965.
Fellow, MRC Clinical Genetics Unit, Institute of Child Health, London and MRC Population Genetics Research Unit 1965–1967.
Lecturer/Consultant in Medical Genetics, QUB 1967–1975.
Professor of Medical Genetics, QUB 1975–2000.
Emeritus Professor in Medical Genetics, QUB 2000–present.
Appointed OBE 2003.
Fellow of the Royal College of Physicians of London.

Fellow of the Royal College of Physicians of Edinburgh.
Fellow of the Royal College of Pathologists.
Fellow of the Faculty of Public Health Medicine.
Member, UK Gene Therapy Advisory Committee 1993–2006.
Chair, UK Gene Therapy Advisory Committee 1996–2006.
Member of UK Human Genetics Advisory Committee 1996–2000 and Commission 2000–2001.
Author of more than three hundred scientific papers, nineteen chapters in books on genetics and the peer publication *The Recognition of Neural Tube Defects* (1983).

# 8

# Professor John Bryant

## Molecular biologist

You don't become a Christian by osmosis, says John Bryant. He should know. He is Professor Emeritus of Cell and Molecular Biology at Exeter University. Since his student days at Cambridge, where he read biology before gaining a PhD in biochemistry, John has been studying cells, in particular the way genes work.

*John, what aspect of cell life do your specialise in?*

I've spent thirty years looking at those two minutes when a cell starts to copy its DNA to make an extra copy so it can divide. How it does that is wonderful, awesome and extremely complicated. We still don't know the full picture.

*When did you first become interested in science?*

I grew up on a council estate near Croydon, just on the edge of South London's green belt. One of my primary school teachers was fascinated with nature, and by the time I was eight, so was I. My parents encouraged my immense curiosity about the living world. We didn't have much money but I was surrounded by love. I also absorbed the faith of my parents, which was very real and very upfront. But I had to work things out for myself and by the time I was a teenager, I had become a Christian.

*What do you say to those who argue that faith and science don't mix?*

First we need to understand that science is a particular way of looking at the world based on observation, measurement and experiment. The data obtained from these tell us about those elements of existence which are open to that sort of approach. Science does not deny the possibility that there are other levels of existence – such as spirituality – or that there is a God.

We can't know Jesus Christ through science but we can know a lot about him through the scientific methods of archaeology. We can verify much of the history and archaeology of the Bible through research. A scientific mind can help you say Jesus did exist. But by itself it doesn't bring anyone into a personal relationship with him.

*As a scientist, how do you view miracles? Can a miracle be rationally explained or scientifically disproved?*

There are different types of miracles. There are occasions when God uses natural phenomena to bring about his purpose – what might loosely be called 'coincidence'. Then there are times when God appears to intervene directly. Things which normally don't happen do happen. Scientific laws aren't prescriptive like the laws of the land. Scientific laws don't say what can't happen, they simply describe what does happen. If we believe in a God who cares for his universe, then it is logical to suggest that there are occasions when he might directly intervene.

*The greatest miracle in the Bible is the resurrection of Jesus. What is the significance of the resurrection for you?*

The resurrection verifies the authenticity of Jesus' divinity. Without the resurrection, Jesus would have been just a good teacher, great philosopher and ethicist. Many of the things Jesus said don't make sense without the resurrection. I agree with C.S. Lewis's analysis that when you read what Jesus said, you have to conclude that he was mad or bad, or that he was who he said he was – the Son of God.

*The virgin birth is another instance of God intervening in human history in a mighty way. As a scientist can you explain the virgin birth?*

No, I can't, either as a scientist or as a Christian. It can't have been a divine cloning experiment or Jesus would have been female. It is a matter of faith that Jesus was born divine and human.

*In our technological age, people increasingly look to science to answer all of humankind's problems. Is science all-powerful?*

No, there are many areas where science doesn't have answers – and not just in the spiritual realm. We don't know, for example, how life began. We can't run a repeat experiment to find out. We will never know for sure and people are happy to accept that. It is interesting that when a scientist says they don't have an answer, the public accepts that. But when a Christian says they don't have an answer, the response is often something like, 'Your faith can't be very strong, then!'

*Does DNA hold all the answers about a person?*

DNA is like a recipe book for building a human being from one cell up to the seventy million million cells in the human body. It provides the biochemical structures which build our cells and determine the way they work, including the biochemical messages that enable the brain to work. But I believe humans are more than the sum of their DNA. Take Paula Radcliffe as an example. She has long legs and a slow pulse rate. Her muscle fibres have slow-twitch predominating. She is genetically predisposed to be a long-distance runner. But DNA alone does not make Paula a top athlete. The support of her parents, her dedication to training, her determination to develop her gifts are not genetic characteristics. God has not programmed us through DNA to be a particular sort of person, even if there certain heritable traits in our character. Environment plays a big part in making us into who we are. That's why, when some people become Christians, they completely change not only in what they do but in how they do things.

*The media sometimes reports that some behaviour is genetically determined – that, for example, there is a selfish gene, a fat gene and a gay gene. Do such genes exist?*

People can have a genetic disposition but a major survey of genes and behaviour published about five years ago suggested that there was only a very small number of cases where we could make a direct link between genetics and behaviour, and even then it wasn't clear-cut.

When Richard Dawkins coined the phrase 'the selfish gene' he wasn't talking, as many people believe, about a gene that makes people selfish but to the idea that the sole purpose of life is to replicate DNA. A gene, therefore, has to be selfish in building a replicating living being. There isn't a gay gene. There may be a very strong predisposition to homosexuality but we can't say it is pure genetics.

*One of the areas in which people are looking to science for the answers is incurable diseases. Does gene therapy hold the longed-for cures?*

There are several thousand conditions we loosely call 'genetic diseases'. In practice, only relatively few could be cured by gene therapy. The reason why there are so few is that if we are going to change the offending gene, we have to find some way of getting into the affected cells. With seventy million million cells in the human body, that is some challenge.

Gene therapy is targeting those diseases where the effect is located in one particular type of cell, such as cystic fibrosis in the lungs and severe combined immunodeficiency disease (SCID) in the bone marrow. We are not changing future generations, we are simply changing the gene in the affected person. I don't have any ethical difficulties with that at all. With cystic fibrosis the results have been very iffy. But the success rate with SCID in Britain is nearly one hundred per cent.

*How does gene therapy work? Do doctors – like car-mechanics – surgically remove a faulty gene and replace it with a healthy one?*

There are different levels of gene therapy. Cystic fibrosis, for example, is a problem with the layers of cells that line particular organs, particularly the gut and lungs. The gut is more difficult but the lungs are accessible, so we can put a correctly functioning gene into a piece of DNA called a vector, which will carry DNA into the cell. This gets into the lungs via an asthma inhaler. With SCID there has to be surgical intervention. The bone marrow cells have to be removed and manipulated. Once the new gene is in place those cells replicate and overtake the malfunctioning cells.

*When scientists approach government for legislation to allow techniques which arouse public disquiet – such as embryonic stem cell or cloning research – one of their arguments is that of finding cures for conditions such as Parkinson's, Huntingdon's and Alzheimer's diseases. Will gene therapy cure such conditions?*

Gene therapy is not a cure-all. A few forms of cancer are genetic and there is a lot of hope there. In the States they've had some success in inhibitory gene therapy, whereby the affected gene is switched off. We could think about gene therapy for Huntingdon's but it would have to be carried in the embryo before it was implanted into the womb. What is tending to happen where such a condition occurs in a family is that they are opting to have test-tube babies so the embryo can be tested before it is implanted. If the embryo has the Huntingdon's gene it is rejected. Currently, on the NHS, there are some seventy conditions which can be tested for in IVF embryos prior to implantation.

*Do you have moral concerns about research on embryonic stem cells?*

In Britain nearly all such research is done on embryos that have been generated by IVF and are not needed to start a pregnancy. Embryonic stem cells are harvested from the blastocyst – which is about the size of a pinhead, about 0.15mm across – at around the six or seven-day stage. In nature, this is just before the establishment of a pregnancy, which is generally completed by the end of the second week after fertilisation. The blastocyst can generate all of the two hundred and thirteen human cell types. Because of the stage of its development, I don't ascribe full human personhood to the blastocyst. In nature, seventy to eighty per cent of blastocysts don't implant into the womb and so don't develop further. So I don't think you are destroying a human person by using the stem cells.

I have two competing opinions about the rights and wrongs of this. On the one hand, I think it is great to be able to use cells to repair people who are in extreme need. On the other, I'm worried that we could be turning the human embryo into a commodity. At the moment I am in favour of embryonic stem cell research under strictly regulated conditions.

*The body governing those regulations, the Human Fertilisation and Embryology Authority, has given scientists permission to insert human DNA into the empty ova of animals such as rabbit or cow. Some people fear this is a step towards scientists making a hybrid. Do you?*

This is not about making hybrids. There are just not enough embryos available for stem cell research. As getting supplies of human eggs is invasive, scientists have been asking for a long time if they could use animal ova as carriers for human genetic material. The idea is not to implant anything but simply to grow human stem cells for research. The amount of animal DNA in the empty egg is twenty out of twenty-six thousand genes – around 0.08 per cent. The technique will allow scientists better to understand human cells. But I would never want to see the results of such experiments used directly in human therapy. We don't know whether the contents of the egg cell or those twenty genes would be carrying anything that might be harmful to humans.

*By responding to human need, scientists show that are not working in a vacuum. But does the science community think that because something is possible, then it should be allowed? In other words, where is the morality in science?*

Science has an ethical framework. I know of only two scientists who would want to clone a human being. When genetic engineering techniques were starting in the 1970s, it was scientists who raised concern about possible misuse, not lay people. Scientists contribute to legal and ethic guidelines. It is for wider society to monitor, reflect and regulate what we require of science.

*John, your research has taken you into the area of genetic modification. When some sections of the tabloid press write stories on GM, they usually focus on food and include words such as 'scare', 'alert', 'contamination' and 'Frankenstein'. Are there less emotive areas where genetic modification is happening?*

Yes, GM is used in the treatment of certain diseases, such as severe combined immunodeficiency immune disease (SCID) via gene therapy. GM is also imbedded in drug manufacture. There are scores of drugs and vaccines for human and veterinary use that are made by genetically modifying bacterial cells. Insulin is probably the best known, because it was the first. There is a whole new generation of products, of which insulin is only one, which lack the harmful side-effects of earlier versions. If in recent years you've had vaccinations before travelling to the tropics, the chances are those vaccines were made by genetic engineering.

*On the subject of GM food, what do scientists mean when they talk about genetically modifying a crop?*

The technique is to take a gene from a cell of one organism and splice it into the genes of another. It is the same basic technique for genetic modification of bacteria, animals, plants and fungi. With more complex organisms, such as mammals and plants, the technology is slightly more complicated than with bacteria but essentially GM is the movement of a gene or a small number of genes from one living organism to another. This is true whether you transcend species barriers or take genes across varieties of the same species. The idea is that the gene you're transferring confers a characteristic you want the recipient gene to have.

Maize, for example, is particularly prone to a number of diseases. In Britain, it is the European corn borer: in Africa, it is the African stem borer which destroys the crop. The bacterium *Bacillus thuringiensis* produces an insecticidal protein. For years, organic farmers have scattered the spores of the bacterium on to crops such as cabbages to stop caterpillars. To make insect-tolerant crops scientists take the toxicity gene from the bacterium and introduce it into the maize. It is completely safe for mammals to eat. When an insect starts to munch on the crop, the toxin stops it.

*Why, then, if GM is being used to bolster food production, is it talked of with words such as 'alert', 'contamination' and 'Frankenstein'?*

GM has acted as a lightning rod for campaigners who are anti-big business. The green lobby and big business have both misled the public about GM food. Big business said that GM will feed the world. No, it won't. It will increase plant-breeding abilities but on its own it will not feed the world.

Green campaigners lied by dressing up in biochemical protection suits to handle GM crops, suggesting to the public that these crops are very dangerous. They are not dangerous. Indeed, in the eleven years that GM crops have been grown commercially in other parts of the world, there have been no problems for human health. Some 'conventional' breeding techniques have involved artificial intervention to produce varieties that could not occur in nature. In my view, whether it's GM or conventional, we should be focusing the application – which genetic trait is being introduced – rather than on the technique. There are, for example, certain applications of plant breeding that as a biologist I am not happy about, whatever technique is used. One is to put herbicide-tolerance – the ability to resist weed-killers – into some domestic rice varieties. The problem is that wild rice, which is very closely related to the rice we eat, is a serious weed. If wild rice were to cross with herbicide-tolerant rice, the result will be a weed which is resistant to weed-killer. Whether the resistance gene is introduced by GM or by conventional methods, that seems to me to be a very silly thing to do.

*In October 2007,* The Guardian *splashed the headline: 'I am creating artificial life, declares US gene pioneer' over an interview with Dr Craig Venter. Is Venter, or any other scientist, really able to create artificial life?*

The headline was misleading. Venter is talking about making a synthetic chromosome. Through genetic mapping and biochemical analysis, we can determine the building blocks in genes. For some organisms we know just a few, for others we know the complete set. It is not difficult to build a gene, but you'd need a robotic machine because it is tedious. If you can build one gene, you can build any number. We know how to get a gene to work so getting a chromosome to work is not a great leap forward. The big question is: Can a living cell be built from nothing? One of the main jobs of DNA to provide recipes for biochemical machines called proteins. But you need proteins to provide recipes for proteins, so you go round in circles. The challenge is: How do we invent a self-replicating system inside the membrane of a cell? It is unlikely that we'll ever generate a living cell from scratch. But it wouldn't disturb my faith as a Christian if somebody did it.

*Craig Venter – who raced Francis Collins to map the entire human genome – is, among others, on record as saying that science and personal faith are incompatible. Are they?*

Science, in this area, is neutral but if you are a scientist who is a Christian, it adds a dimension to your work. I'm not saying that Richard Dawkins doesn't get joy from scientific discovery because I know he does. I know that Dawkins is awestruck by the beauty and the complexity of nature. But as a Christian I have a sense of seeing the work of the Creator. When my team and I make a new discovery I'm overwhelmed by the wonder that, apart from the few of us, only God has seen what we've seen. For me, knowing about the works of the Creator is a form of worship.

*What about when society worships science? How comfortable are you with that idea?*

Some people do view science as being the only approach to understanding the universe. There are other inaccurate views of science. Science is used as a validator of products, pseudo-scientific words are used to add authority to beauty products. The word 'science' simply means 'knowledge'. Then there's the post-modernist strand that says scientific findings are just as prone to doubt as any sociological theory and that the findings are true only if I accept them. And some people are suspicious of technology.

*John, if someone asked you as a scientist to prove God, what would you say?*

I wouldn't even try. Science has its limits, which are based on the physical make-up of the universe. As a Christian, I would say that there are elements of life which can't be scrutinised or explained by the scientific method of observation, measurement and experiment, and those elements include the spiritual. We can't scientifically put God to the test but we can experience him in our lives. And we investigate the life and claims of Jesus.

*To what extent can science help us to understand how prayer works?*

I think it's beyond science to tell us whether or how prayer works. I know that when I pray, I put myself into God's presence in a way that's different from when I'm at work in my lab. Over the years, scientists have talked about prayer as the engagement of the time-limited with the timeless. I don't know how the mechanics work but I do know prayer works.

*One of reasons why some people don't pray is that they struggle with the image Christians paint of God as all-loving and all-powerful, yet also as one who doesn't always intervene to stop natural disasters or personal tragedies. How do you picture God?*

I believe God is all-loving and all-powerful. But although he is all-powerful, God created a universe that is open, not cut and dried. I don't know why an all-loving God doesn't intervene to stop suffering but I do believe that this world isn't all there is, and the answers may be revealed to us in another place, at another time.

*Why is it important to you that God created the universe and that it didn't come into being some other way?*

I believe that God is the Creator and Sustainer of the universe. For me, God creating the universe and everything and everyone in it is a given. If there were no God, the universe would be completely inexplicable. Even with God, it is pretty mysterious!

When the Big Bang hypothesis was first put forward, many Christians thought: 'Great, science agrees with us. There was a creative act that started the universe.' I first came across the theory of evolution when I was eleven and have never found it incompatible with the Christian faith. Evolution theory itself neither eliminates nor includes God. It merely says this is a set of mechanisms. The big question remains: Who first generated the evolutionary process? To me, the answer has to be God.

## FACT FILE

*Professor John Bryant*

BA Natural Sciences (Botany), Cambridge.

PhD Plant Biochemistry, Cambridge.

Fellow of the Wessex Institute of Technology 2003–present.

Professor of Cell and Molecular Biology, University of Exeter 1985–2001.

Professor Emeritus of Cell and Molecular Biology, University of Exeter 2001–present.

Visiting Professor at West Virginia State University, USA 1999–2007.

President of the Society for Experimental Biology 2003–2005.

Chair of Christians in Science 2001–2007.

Co-Chair of the Higher Education Academy's Special Interest Group on Teaching Ethics to Bioscience Students 2002–present.

Author of around one hundred scientific papers and of books, including *Bioethics for Scientists* (with John Searle and Linda Baggott la Velle, Bognor Regis: John Wiley & Sons, 2002), *Life in Our Hands – A Christian Perspective on Genetics and Cloning* (with John Searle, Nottingham: Inter-Varsity Press, 2004) and *The Eukaryotic Cell Cycle* (with Dennis Francis, Oxford: Taylor & Francis Ltd. 2007).

Broadcaster.

# 9

# Dr Denis Alexander

## Immunologist

To the layperson even the titles of some of Denis Alexander's research papers make no sense. 'DNA Damage-induced Bcl-$x_L$ Deamidation is Mediated by NHE-1 Antiport Regulated Intracellular pH' sounds more like the subject of a bonus round on *University Challenge* than something that actually does any good. Yet the reality of Dr Alexander's work – to find a cure for cancer – is something to which everyone can relate.

Having previously worked for the Imperial Cancer Research Fund (now Cancer Research UK) and been Associate Professor of Biochemistry at the American University in Beirut, Denis has, since 1989, been supervising research groups at the Babraham Institute, Cambridge. In 2006 he became the inaugural Director of the Faraday Institute for Science and Religion, St Edmund's College, Cambridge

*Denis, what does your current work entail?*

I work in immunology, looking at how the body defends itself against viruses and bacteria. My team and I are looking at the very early stages of cancer, particularly the cancers of the immune system. We want to know what makes cells proliferate out of control. We are not working directly on HIV, but we are working on the white blood cells known as T-lymphocytes, which are the ones that get infected by the HIV virus. We're also looking at cell damage. Cells get DNA damage all the time. When they do they normally die – each cell of the body is programmed to commit suicide if it is damaged. It is part of the body's defensive mechanism against getting cancer. Cells are dying all the time in our bodies but we don't realise it.

Occasionally damaged cells don't switch on the suicide programme. Instead they continue dividing out of control with more and more DNA damage, and that's how some cancers get started. This is a case of where life – damaged live cells – is more dangerous than death. We are trying to work out how those cells stay alive, even though they are damaged, and how to switch their suicide programme on again so that they don't develop into cancer cells.

*What, or who, inspired you to get into science?*

I suppose it was a gradual process. My mother read physiology at Oxford and clearly had a great influence on me. My father was a timber merchant and was very interested in natural history. We lived close to the countryside, so I spent my youth roaming the fields and woods in the area. At school I had a very good biology teacher. It grew from there. I went to Oxford and read biochemistry, and then did a doctorate in neurochemistry at the Institute of Psychiatry in London.

I've always been fascinated with how the world works. Obviously it is a huge challenge to find out how cancer works – and it has serious implications – but I've always had a great interest in cancer research. Even when I was at school I thought: Wouldn't it be fantastic to tackle the problem of cancer?

*What about your faith? How did that start?*

I was nurtured in the Christian faith by my parents. But I made my personal commitment at a Crusaders' camp when I was thirteen. Crusaders was an interdenominational Bible group. Every year we used to go from our local branch in Banstead, Surrey, to a house party on the Isle of Wight. It was during one of those occasions that I put my trust in Christ as Saviour. It was a very clear and conscious decision on my part. I realised that, although I had been brought up in a Christian family, it didn't make me a Christian. I knew I had to personally commit my life to Christ.

*Historically faith and science have sometimes been seen to be in conflict. Are faith and science incompatible?*

I've never had a conflict between being a Christian and a scientist. For me, understanding how the world works at a scientific level has always been part of my worship. But it's not just an absence of conflict. There are a number of similarities between personal faith and science. For example, the process of exploring the reasons for putting one's trust in God is very similar to scientific inquiry. The scientific and religious quests are searching for coherence – we want to make sense of lots of different kinds of data. So in a laboratory, we try to make sense of all the pieces of data that come from observation and experiment. We are looking for some story that will make sense of all those pieces of the puzzle. This is what scientists mean when they talk about a 'theory' or 'model'. It's like having a map in your hands that makes the world around you scientifically coherent. The religious quest is a similar pilgrimage – we are looking for what makes the best sense of the universe that we live in.

Another similarity would be what philosophers call falsifiability. Karl Popper said that science was defined by what you can exclude or falsify. His favourite example was: If you believe that all swans are white and you write a scientific law that says so but then you go to Australia and see a black swan, the theory is blown apart. It doesn't matter how many white swans you keep counting, one black swan means your theory's gone. In the religious field, Christianity, Islam and Judaism depend on historical data. Christianity, for example, depends upon being able to investigate Jesus as a real, historical figure. In his first letter to the church at Corinth, Paul uses the same argument about Jesus: 'If Christ has not been raised, then our faith is in vain.' Paul's faith could readily have been falsified by someone producing the embalmed body of Jesus at that moment, but of course that didn't happen. In this instance, the scientific and religious ways of thinking happen to be very similar.

*Some people think faith that is based on emotionalism while science is grounded in rationalism – and never the twain shall meet. Does science run only on what the senses tell us or is it partly built on things unseen? In other words, do scientists need faith to do science?*

Yes, in a general sense they do. The word 'faith' can be used in different ways but scientists often don't realise how much implicit faith they have. For example, a scientist has to believe that the universe is rational and reproducible – that it has properties which can be investigated in a rational way; scientists have to believe that if they do an experiment this week, it will produce the same results next week. There are all kinds of implicit background understanding or 'faith' that all scientists have, whether they realise it or not.

Looking more broadly, we can see the influence of Christianity on science. If, for example, the Babylonian creation mythologies had been the mythologies that had dominated in Europe for thousands of years, it is very unlikely that we would have seen science develop. In

Babylonian polytheism, different gods were fighting over different bits of the world. A scientist wouldn't know where they were, because if a god changed their mind – and the Babylonian gods were fickle – then why would you think that the properties of the world were reproducible and therefore worth investigating?

The Judaeo-Christian image of a Creator God provides the basis for the idea of order in the universe. From order we have the idea of laws describing how the universe works. Order means that if the law worked yesterday, it should hold true tomorrow. The idea of laws has strong Christian underpinning, which can be seen in the writings of Sir Isaac Newton, René Descartes and Robert Boyle among others, who clearly saw that if there were religious laws for the universe, there should also be a parallel set of scientific laws which could be discovered. That is now central to our scientific enterprise and understanding, but many scientists don't realise the religious roots of the scientific law concept.

*Do the similarities between faith and science include the desire to reach out to something greater?*

Scientists are in a privileged position. Because they spend all their time studying this wonderful creation, scientists should realise a desire to reach out to something more than anyone. But we don't have to be scientists to recognise God in creation. Anyone who looks at the beauty and wonder of the world should pick up hints that there is something more to life than just what they see – a hint that there is some Being who is behind it all and who makes sense of it all. A woman once told me that after she'd looked at the spring flowers in her garden it dawned on her that there must be a God – that with all the beauty around, it was ridiculous for her to be an atheist.

*The Bible describes the pinnacle of creation – humankind – as being made in 'the image of God'. What does being 'made in the image of God' mean to you?*

There are many aspects to humankind being made in the image of God. For me, it speaks about relationship. Humankind can have a relationship with God the Father, God the Son and God the Holy Spirit. Relationship is rooted in the very character of God, so being made in his image means reflecting something of that character. God has called humankind into fellowship with him. Bestowing his image on humankind is a way of expressing the fact that it is only humankind who has the possibility of knowing God in a personal way. It is only humankind who can pray or who can hear God's voice. Being made in God's image also means responsibility. God has made us responsible for the care of our planet. We cannot treat it thoughtlessly.

*On the other side of the equation, what image do you have of God?*

Because I was brought up in a Christian environment, my picture of God has been shaped by the Scriptures – he is a personal God who has come into the world in the person of Jesus to save us. Without this revelation, I might conclude that the world was created or designed by a superior power or intelligence but I wouldn't know whether that power was abstract or personal, kindly or malevolent. Nature alone doesn't tell us that.

*You have devoted your professional life to understanding cancer and finding cures. Every day you confront the problem of suffering. Why, do you think, does an all-powerful God allows people to suffer?*

Ultimately only God knows why there is suffering. Some suffering comes about because of human evil. People are free to behave as they will and sometimes people make harmful choices. As a biologist, I see that life as God created it is a package deal. Life on Earth, and probably in the whole of the universe, is based on carbon. With carbon-based life you have to have transience, you have to have life and death. It is built into the system. Life without death makes no sense – that's the package deal. Also, if you're going to have different biological organisms there must be variation. (Without variation all humans would be one great clone.) We look different and have different personalities because of variation in the genetic alphabet. Everybody in the world differs by roughly one in every thousand letters in the human genome (the genome means all the information contained in our DNA). The result is we are all different and that helps us to understand how God loves us as individuals. Fantastic!

The downside is that once there is genetic variation, there is disease, because genetic variation sometimes causes disease – including cancer. So suffering in the form of disease seems to be the necessary cost of living in a carbon-based world. Imagine, though, the alternative – a world where nobody died. The world would be packed – standing room only! Clearly, when God made our planet's set of carbon-based living organisms, he did not intend that we should live and reproduce for ever and never die. The Bible gives us another perspective in that God has a new Heaven and a new Earth in mind for us. Only against the backdrop of a new Heaven and a new Earth can suffering start to make sense.

*You are the Director of the Faraday Institute for Science and Religion. One of the institute's aims is to 'provide accurate information on science and religion for the international media and wider public'. With the advance of scientific techniques, particularly in medicine and genetics, scientists are sometimes accused of playing God. What procedures are there to keep scientists in line with what the wider public wants?*

Scientists in the UK are very tightly governed by law, government regulations and ethics committees. For example, at the Babraham Institute we have an ethics committee. If there is anything that raises particular ethical questions, it has to go to the committee for consideration and permission. No scientist can do anything in the laboratory that is deemed to be unethical. Scientists are also governed by funding bodies. For example, when you apply for a research grant you are asked to explain any ethical issues raised by the proposed research. So scientists are forced to consider ethical questions before they get money for research. There is also a measure of self-regulation. Scientists know that they have to be honest in their work. If a scientist makes false claims or tries something that is unapproved, then that is likely to be the end of their career.

*To a layperson a theory is often regarded as an idea that can be readily dismissed, as in 'It's only a theory.' What do scientists mean when they talk about a theory? Do they mean something that is cast in stone?*

When scientists use the word 'theory' they take it to mean 'It is something I do believe'. They don't use it in the popular sense of 'It's only a theory', meaning 'I can choose to believe it or not'. For a scientist, a theory is a map on which data fits and makes sense of the world. So, for example, the theory of Big Bang cosmology puts together lots of data that actually make a lot of sense if this universe actually had a beginning. That's why scientists get very passionate about their theories and will defend them very strongly.

*One of the things that some Christians get passionate about is Darwin's theory of evolution, which differs from the biblical account of God creating the world in six days. How do you read the early chapters of Genesis – as literal truth or as a metaphor containing truth?*

To me, 'literal' means 'what the author originally intended to say'. The challenge of the Bible is to find out what the author did originally intend to say, given that the Bible contains different kinds of literature. For example, if we said: 'So-and-so has really got a chip on their shoulder', the literal interpretation of that phrase – in the sense of what is meant – is that 'this person gets upset easily'. Language is full of figurative statements and metaphors like that.

So how should we approach Genesis? Writing in the fifth century, Augustine offered an interpretation that wasn't influenced by scientific theories, because he was writing in the days before science. We might think that in his commentary *The Literal Interpretation of Genesis*, he argued that creation took six 24-hour periods. But he didn't. He saw the early chapters of Genesis as being figurative.

The third-century theologian Origen also had a very figurative understanding of Genesis. This suggests to me that for centuries the early chapters have been treated by many commentators as a figurative, theological text inspired by God, foundational to the rest of Scripture, but not intended to be regarded as a scientific text. It's a pity that in the twentieth century people started reading Genesis as though it were a scientific textbook. Trying to impose science on the text is a misuse of Scripture.

*Some scientists interpret the world this way: God created the universe through Big Bang and continues to sustain it through evolution. Is this the way you see it?*

In general, yes. But for me 'creation' not only speaks about beginnings, but also about on-goings and the future. God clearly brought the universe into being at the beginning, but it is by no means static. God is an active Creator who is creating all the time. We live in a dynamic universe. Old stars die as new stars come into being, for example. We have an active God who is bringing things into being and making other things go away.

*Another theory used to describe how life came about is Intelligent Design. How credible is that model?*

Different people use the term Intelligent Design [ID] in different ways but generally it is used of an anti-Darwinian movement that began in the early 1990s in America. It was spearheaded by Phillip E. Johnson, a Professor of Law at the University of California. Johnson was concerned about what he saw as naturalistic philosophy – explanations for life that didn't include God – invading the scientific community. He saw Darwinism as an icon of naturalism within the scientific community, especially biology. So he launched a counter-attack in the form of ID.

ID is the idea that there are certain entities in biology that are so complicated that they couldn't have come about by the gradual incremental process that Darwin suggested. The most often cited example is the bacterial flagellum, a tail-like object which protrudes like a little oar with an outboard motor from the body of a cell. It has some thirty to forty components, making it very complex. ID supporters claim that evolution cannot explain the existence of something so complex. They argue that something that is irreducibly complex (that is, if one part were taken away it won't work) must have been designed, therefore there must be a designer. They are also very concerned to say that they are not religious, and that they are not starting with the Bible and trying to make the science fit, but are starting just with what is.

My own view is that ID is a form of the old God-of-the-gaps argument, where God is used as the explanation for those questions science can't answer. The problem with that argument is that as science answers more questions, God becomes more redundant. The only difference with ID is that you could call it a 'designer-of-the-gaps' argument.

Since Johnson and the ID biochemist Michael Behe championed the flagellum as evidence of ID, we now know that it contains different modules that can have quite different functions in other bacteria. For example, there's a module in the flagellum which other bacteria use to inject poisons into each other for self-defence. So the ID argument doesn't really work because it turns out that the flagellum is not 'irreducibly complex' after all. It contains parts that evolved independently and incrementally, each giving an advantage to the organism that possesses them. Evolution doesn't work by bringing everything into being all at once, but bit by bit.

*Does the idea of incremental evolution rule out the possibility of God as Designer?*

No. The classic argument for design says that God is a Designer who brings into being all the properties of the universe without exception. In other words, it focuses on what we do understand and have access to. Of course, you don't have to be a scientist to understand design in the universe. Have a look in your garden or take a walk in a park. The design we see in the wonders of nature ultimately reflects the mind of the Designer.

*If nature reflects the mind of a Designer, would you point to nature as proof that there is a God?*

As proof, no. I don't think you can prove God like that. But on a walk in the countryside on a beautiful spring day I would hope that anyone could see in the beauty of creation hints of something beyond. A look at the night sky would get me thinking about how the universe is finely tuned so there can be life on Earth. It doesn't prove God but it hints that there is something going on worth investigating.

Then there's the moral life of human beings – the fact that deep down everyone believes in justice. We all believe there is a sense of right and wrong, even if we differ about what we think is right and wrong. To me, this suggests that somewhere there is an ultimate moral arbiter of things.

The fact that we exist also fits with the idea that there is a God. Why else should we complex human beings have wonderful poetry, music, drama and art, and be people with a sense of justice and exist on this planet, if there is no God? It is the existence of a God who has purposes and intentions for the world that makes best sense of this on-going drama of life in which we find ourselves.

The ultimate piece of evidence is Jesus. The biblical account of his life – what he said, what he did, what he stood for – invites, and stands up to, investigation. Jesus is still changing people's lives. When I was working in Beirut during the Lebanese civil war, I met a woman from a Druze background who had become a Christian. The Druze sect is an offshoot of Islam and it is rare for someone to change faith. I was intrigued as to why she had become a Christian.

She said she became a Christian because the Druze faith had no answer to the problem of suffering. It couldn't help her come to terms with the horrors and bloodshed of the civil war. 'Then,' she said, 'I heard about the cross of Christ and how Jesus had suffered and what his suffering and death did for the world. That resonated with me. I suddenly saw that here was a God who identified with the evil of humankind. Here was a God who knew what suffering was. Here was a God that I could relate to.'

## FACT FILE

*Dr Denis Alexander*

BA (Biochemistry), Oxford.
PhD in Neurochemistry, Institute of Psychiatry, London.
Fellow St. Edmund's College, Cambridge 1998.
Associate Professor, Human Genetics, American University of Beirut Medical Faculty, Lebanon 1981–1986.
Research Fellow, Imperial Cancer Research Fund, London 1986–1989.
Chairman of the Molecular Immunology Programme and Head of the Laboratory of Lymphocyte Signalling and Development, The Babraham Institute, Cambridge 1989–2005.
Senior Affiliated Scientist at The Babraham Institute, Cambridge 2006–present.
Director of the Faraday Institute for Science and Religion, St. Edmund's College 2006–present.
Editor *Science & Christian Belief* 1992–present.
Committee member, Christians in Science 1988–present.
Fellow, International Society for Science and Religion 2002–present.
Lecturer, writer and broadcaster on science and religion, whose latest book is *Creation or Evolution – Do We Have to Choose?* (Oxford: Monarch, 2008).

# 10

# Dr Christopher Southgate

## Biochemist

When a scientist discovers faith, it is easy to imagine that it was while gazing at far-distant galaxies or peering at intricate cell structures. While such eureka moments undoubtedly come to some, that is not the way it was for Cambridge-educated enzyme inhibitors specialist Chris Southgate. God spoke to him through the heart.

After he married, Chris turned his back on a scientific career and became a house-husband. He looked after his stepson, Jac, and tried his hand at writing. The family settled on the northern fringe of Dartmoor – mother of rivers, mover of men – where Chris developed his poetry-writing, much of it inspired by the granite-cropped wilderness. He took an ordination course as a lay candidate, became pastoral assistant at Exeter University and in 1993 joined the teaching staff as an honorary fellow. His specialist subject is faith and science.

*Chris, how did you come to faith?*

Sandy, the woman I loved, was ill. I was in England. She was in America. I couldn't help her. I was twenty-six and, with the confidence of that age, I thought I could fix anything. But here was something I patently couldn't fix. All I could think to do was pray. I hadn't willingly gone to church regularly since I was eight. There had been one or two moments in my teens when I thought about God, maybe through a particular piece of music or a poem, but it seemed to be as if through a half-opened door and I never did anything about it.

God answered that prayer. I remember that at the time I had a gut sense of assurance that everything would be all right. (We have now been married for twenty-six years.) That was only the start of the journey.

There were several months of exploration. I started to read the New Testament, talked with a few people and started going to the chaplaincy at the university where I was working. I'd always thought that Christianity was about keeping rules. I was wrong. I discovered that faith is about freedom – the freedom that comes from laying down the baggage of the past and being able to go forward. I came to see the gospel as radical. It affects all sorts of issues in your life. I was hugely challenged. Over the months I realised I was developing a relationship with God. Nobody was pushing me, it was a gradual and solitary process. I was never a beaming-smiles Christian but I certainly knew every part of my life had been changed.

*What areas of faith and science do you specialise in?*

Key topics include how the universe began. Did it just begin all by itself or did God set it up? Is Big Bang cosmology a hint of God at work as the Creator and the Designer of the universe? If we understand ourselves as evolved animals of a particular type, where does God fit in? Where does the teaching on humans being special to God fit into the evolutionary model?

My background in biochemistry increases my sense of wonder about the world, rather than gives me the sense that all the questions have been answered. I tell my students there are no knock-down answers. They must take their own positions which they need to be able to hold with integrity and defend intellectually. The question that gave them most trouble last time I taught was the one I explore in my latest book, *The Groaning of Creation* – how could a good God have used a process like evolution, full of struggle and suffering, to create?

*And the answer is?*

There are never any neat answers to the problem of suffering. But you should read the book!

*And how should we read the Bible? Are its early chapters a scientific digest of how the world came into being?*

I have read many scientific textbooks. The Bible is not a scientific textbook. It talks about the nature of God's relationship to the world and to human beings in a very different way from the way science does. The biblical account of creation and the theory of the Big Bang are two descriptions of how the world came into being. They are not entirely separate. They interact and inform each other at key points. There is no reason to take the Bible so literally that science knocks it down. That is not a helpful way to work.

*Which part of science, then, would knock down the idea of God literally forming Adam out of dust?*

Cosmological physics gives us a plausible and coherent map of the history of the universe which goes back about thirteen billion years, almost to the Big Bang. Evolutionary biology gives us a history of living beings on earth that goes back three-and-a-half billion years. It tells us we are not made of dust. We are made of cells which contain processes that are related to the processes that began all those billions of years ago. One of the reasons humans are the way we are is because we are evolved animals. But where some scientists make a mistake is to say human beings are nothing but that.

Humans are much more than just evolved animals. We alone, for example, have the capacity freely to choose a relationship with God. Adam and Eve tell us that human nature includes the capacity to reject an awareness of God's wise sovereignty.

*If not everything in the Bible is scientifically provable, what points to it being true?*

In order to prove something, scientists speak in terms of it being coherent, comprehensive, agreeing with known data and fruitful. The Bible is the story of particular people's walk with God. There is a certain amount of external historical evidence which supports the existence of those communities and that they had some sort of story. But you can't prove the truths of the Bible from outside. You have to read it, go inside the story and see whether in your judgement it is, in the words of scientists, coherent, comprehensive and fruitful. People have been doing that for two thousand years and they've discovered the Bible is an amazingly rich account of God's story with human beings.

*Does the scientific process similarly require scientists to take a leap of faith in order to do science?*

In order to explore, a scientist has to take certain things on trust. The whole of science rests on the conviction that there are reproducible patterns going on. Scientific laws and theories are built on trust. We have to take on trust, for example, that mathematics is something to do with the way the world is.

*What can science offer when it comes to the religious idea of miracle?*

Theologically, a miracle is a unique event which particularly displays the character of God in relation to the world. I don't think God goes around breaking the natural laws he has made. That's not to say that he may not very occasionally do so for his own good purposes but I think, in our understanding of his character, God is faithful to those laws and those laws enable us to be what we are.

Science, however, is very bad at assessing the character of unique events. It likes to form conclusions from a consistent pattern of events. Also it has no language for talking about the character and purposes of God. So science struggles to deal with the unique events that I'd claim to be miracles. But science is changing. Two hundred years ago it was thought that, by building a picture of the universe at any one moment, science could predict everything that would happen. Now, through the likes of Chaos theory, physics tells us the universe unfolds in a much more complicated and unpredictable way than that. We cannot determine the future by studying the patterns of the past. Scientific thinking allows for very unusual things to happen.

*As one who is trained to investigate, reason and explain, do you struggle with the idea of miracle?*

No. The world science describes is an extraordinary one. It is a world with a very long history and if anything had been different at any stage in that history then we probably wouldn't be here. So the fact that we are here and live on such a beautiful and astonishing planet is amazing, and science itself is a great source of wonder.

We have an incomplete understanding of the world's natural and spiritual processes. What happens at any one moment is a very complex interaction of things. Sometimes we get a better understanding by running a scientific experiment, sometimes by spending time in prayer. The resurrection of Jesus is the greatest and most striking miracle of all. It was the one time when God not so much broke the natural process as transformed it.

*What convinces you that the resurrection is true?*

Look at the way the lives of the disciples were transformed. Look how the Holy Spirit worked within that transformed community. Look at how we are still in touch with that way of God's working. I don't think the transformation in the disciples can be explained in any other way, other than that Jesus was raised from the dead.

*To many people, human life is a miracle. To some, the gift of a child is miraculous. Science – through IVF – has produced a solution for some people. Scientists are now looking at gene therapy and human cloning for possible cures for other conditions. Do you welcome gene technology?*

Science is very curious to see round the next corner. It doesn't give us the wisdom to know whether this particular corner is one we should be looking round, nor whether the technologies that result from looking round the corner are things that we should do. I am opposed to human cloning technologies which would make clones of other people. My objections are in terms of the risks of the techniques, the possible consequences to the failed clones, the long-term health of the cloned person and the relationship between the clone and its parent. If we think that relationships are particularly important and in some sense God-given, yet we want to create a form of relationship that has never existed before and could not exist but for our scientific ingenuity, then that is questionable.

*What about being able to eradicate debilitating illnesses through so-called 'therapeutic cloning' – isn't that an instance of science benefiting humankind?*

A friend of mine has Parkinson's disease. I want to think that anything that can be done should be done. But I am uncomfortable with therapeutic cloning because embryos, potential babies, are produced only and specifically as a research or clinical tool. So I was really pleased to learn that the latest techniques now being developed may avoid the production of anything that could genuinely be regarded as an embryo.

I respect the integrity of those who say that full human status begins at conception. But it is not a view that I take myself. I see the attributes of human personhood gathering gradually as the foetus develops. Biology only gives us a gradual picture. From the point of view of law and ethics we have to make cut-offs and they are essentially arbitrary.

*What does your Christian faith give you?*

My faith is the reason my life has any meaning and isn't thrashing around in an ocean of confusion and conflicting desires. It is what gives me a sense of forgiveness and a new start from moment to moment. It is what enables me to try to relate to people I don't particularly like and would otherwise definitely avoid. Jesus to me is the Lord we know from the New Testament – the One sent by God to show us what it is to be human and to suffer for us. In science you explore something you regard as passive – just an object of experiment. Faith, though, is exploring – and enjoying – a personal relationship with God, the Creator of the universe.

## FACT FILE

*Dr Christopher Southgate*
BSc (Natural Sciences), Cambridge.
PhD (Biochemistry), Cambridge.
Post-doctoral Fellow at Universities of North Carolina and Bath.
Househusband 1981–1990.

Trained as lay candidate on Anglican ordination course 1987–1990.

Pastoral assistant, Chaplaincy, Exeter University 1990–1996.

Part-time Lecturer in Theology, Exeter University 1993–present.

Chris has also worked as a mental health chaplain, and now is involved in the training of Anglican and Methodist ministers. He is also a researcher on the use of the Bible in environmental ethics, and on the origin of life.

Poet whose fifth collection, *Easing the Gravity Field: Poems of Science and Love* (Nottingham: Shoestring Press, 2006), is the latest.

Editor of *God, Humanity and the Cosmos* (London and New York: T&T Clark, 1999, expanded 2005) and *The Groaning of Creation: God, Evolution and the Problem of Evil* (Louisville, Kentucky: Westminster/John Knox Press, forthcoming 2008).

# 11

# Professor Sir John Houghton, CBE

## Physicist

When we want to know what tomorrow's weather is going to be like, we catch the forecast. When prime ministers want to know the long-range outlook, they turn to Sir John Houghton. For years he has seen which way the wind in blowing. He gave his first lecture on increased carbon levels in the atmosphere in 1967 and has kept a weather eye on them ever since. After studying physics at Oxford – he entered the university at the tender age of sixteen – Sir John got involved in NASA's space programme. In the 1970s, he and a colleague devised a number of weather-measuring devices, which hitched a ride on American satellites. In 1983 he left Oxford, where he was Professor of Atmospheric Physics, to become director general of the Meteorological Office. He had known about global warming since the 1970s. Twenty years later, others were beginning to want to know. In 1990 Prime Minister Margaret Thatcher invited him to address the Cabinet at No 10. Today his prognosis is bleak.

*Sir John, how bad is climate change?*

We are raping planet Earth. Unless we cut back on carbon emissions, floods and drought will increase, there will be less fertile soil for cultivation and millions of people will suffer. The signs have been there for some time. In the 1970s satellites gave us revolutionary new ways of observing and measuring the global atmosphere. The other great revolution for meteorologists was computers. It was one thing having masses of new data, but computers helped us to model and predict climate patterns around the world. In 1988 Margaret Thatcher gave the address at the Royal Society's annual dinner. In it she mentioned the greenhouse effect and how global warming might be a problem for the world. She supported the setting up of the Intergovernmental Panel on Climate Change (IPCC), which included a group of climate scientists from all over the world. I presented the panel's report on climate change to the Cabinet.

Two years later the Earth Summit took place in Rio. The delegates considered our report and every nation signed up to the climate convention, which says that although we are not sure about all the science, we know enough to recognise that we should start to reduce our greenhouse gas emissions. Its key target was for the developed world – those who had most benefited from burning great quantities of carbon fuels for two hundred years – to peg its emission levels to the 1990 levels by the year 2000. The UK managed to achieve this but most nations didn't.

President Bush (Senior) signed the convention for the United States and the Senate ratified it unanimously. Then Exxon and other American energy companies got together and started a massive misinformation campaign. They spent a lot of money putting out the word that the Earth Summit was just a bunch of environmental scientists – that the science was 'junk science' and Americans didn't need to take any notice of global warming. This was despite the fact that some of the best climate scientists in the world are from the United States. Although in 1997 the Clinton-Gore administration signed up to the Kyoto Protocol to reduce greenhouse gases, by the time it came to be ratified the mis-informers had got their man – George W. Bush – into the White House and so the most industrialised nation in the world has not formally committed itself to reducing carbon emissions.

*Before looking at the evidence that global warming is a reality and not 'junk science', could you clarify what is meant by 'global warming'?*

Gases in the atmosphere – the most important of which are water vapour, carbon dioxide and methane – help keep the Earth warm. They absorb some of the heat from infra-red radiation that comes from the Earth's surface when it is heated by the Sun. If it weren't for these gases, the heat would pass through into space. These 'greenhouse gases' act as a blanket and keep us warm. Greenhouse gases have been there for billions of years and are very good for us. If they weren't there, the Earth would be an average of 20°C colder, covered by ice and life would be much more difficult. But the thicker the blanket gets, the warmer the Earth becomes. Since the Industrial Revolution, carbon dioxide levels in the atmosphere have risen by thirty or forty per cent, as a result of us burning fossil fuels, such as oil, coal and gas. Not only is the Earth getting warmer but it is doing so more rapidly than in the past. As the Earth gets warmer, the climate changes.

*Not everyone is convinced that climate change is a result of human activity such as coal-fired power stations or gas-guzzling vehicles. What evidence do you have that climate change is a reality?*

Climate possesses a large natural variability. So identifying clear temperature trends is very difficult. Other factors, too, influence climate change, for example volcanoes, which put dust into the atmosphere. The Sun's radiation has changed a little over the years. Again, it's hard to measure but it is believed that changes in the Sun influence our climate. The scientists working on the IPCC in the early 1990s believed the basic physics of greenhouse gases and that the Earth was getting warmer. By 1995, more warning signs were discerned and the IPCC concluded that there had been around a 0.5°C rise in the average surface temperature of the Earth since the late nineteenth century.

*In 2007 the IPCC was joint winner of the Nobel Peace Prize. In his acceptance speech, its Chairman R.K. Pachauri referred to a clause in the UN General Assembly's founding resolution for the IPCC: 'Noting with concern that the emerging evidence indicates that continued growth in atmospheric concentrations of "greenhouse" gases could produce global warming with an eventual rise in sea levels, the effects of which could be disastrous for mankind if timely steps are not taken at all levels.'*

*Sir John, just how 'disastrous for mankind' will global warming be, if we do not act swiftly or efficiently enough?*

The rate of climate change will be greater than anything that happened in the past ten thousand years. It will be very hard for ecosystems and humans to adapt to that rate of change. Many animal species will become extinct. The first big impact will be increased flooding. As the oceans warm, the water expands and sea level rises. Glaciers also melt, and some of them are melting fast. There may be some melting of the polar ice caps. We are talking about a rise in sea level this century of between 0.5 and 1 metre. That is serious. In the UK, that will cause problems in low-lying areas such as East

Anglia. In London, the Thames Barrier will have to be rebuilt.

On a wider scale, a lot of the world's population lives close to the oceans. In Bangladesh, ten million people live below the 1m contour. Where will they go? In Southern China, twenty-five million people live below the 1m contour. Millions more live on low-lying islands of the Pacific and Indian Oceans. Then there are delta regions, such as the Nile and the Mississippi Deltas. Floods create refugees. There will be a mass movement of people away from these areas. That will have social and financial implications for other countries.

The second big impact is that there will be more heat-waves. Land temperatures will rise by an average of 4 to 5°C. The European heatwave of 2003, which killed at least thirty thousand people, was largely caused by the increase in greenhouse gases. Places such as the Mediterranean and the Tropics will get very warm and dry. We have had a modest increase in temperature of around 0.75°C world-wide. But if we do nothing, by mid-century temperatures will rise by 2 or 3°C, and by 3 or 4°C by 2100.

The third impact will be seen in the hydrological cycle. As the Earth gets warmer, more water vapour will evaporate from the ocean surface. Increased water vapour means increased latent heat and increased latent heat means increased atmospheric energy. So there will be more and heavier rainfall. Increased evaporation will also mean that soil becomes more barren. Water is a very important resource. In some parts of the world it is getting very scarce. In many places it is being pumped out from underground at an unsustainable rate. Wars could well be started over access to water because most rivers of the world are shared by different nations.

Climate change means more floods, more droughts and less fertile soil to feed billions of displaced people.

Floods and drought cause more deaths and commercial loss than anything else. They are the biggest disasters the world knows. Currently, according to research carried out at the Hadley Centre – the UK's centre for climate change research – about two per cent of the world's land area is already affected by drought. Twenty years ago, it was one per cent. By 2050 it is likely to be ten to twelve per cent and thirty per cent by the end of the century. Imagine the impact of that!

*China is reportedly building around one coal-fired power station a week. Why should we in the West be cutting our carbon emissions when developing nations seem determined to increase theirs?*

We in the West have a moral duty. For the past two hundred years, we've grown prosperous on the back of burning fossil fuels, yet the potential devastation will fall disproportionately on the poorer countries. That is not right. There is a tremendous moral imperative for us in the West to cut our emissions very quickly and to use our wealth and skills to help emerging economies develop sustainability.

*How did you come to faith?*

I grew up in a Christian family. I was brought up on the Bible but my faith came alive while studying at Oxford. I went to university just after the Second World War. I was only sixteen and it was full of ex-servicemen who were older and much more worldlywise than me. Members of the Christian Union befriended me. I challenged myself: 'Am I going to put my lot in with these lovely Christian folk, or am I going to go my own way?' In the end the decision was clear – I should throw my lot in with the Christian faith.

*To what extent are your green credentials, part and parcel of your faith?*

As a Christian I believe humankind has been entrusted with this planet. Adam and Eve were told to care for the Garden of Eden on behalf of its Maker – God. They were to be servants, not exploiters. Throughout the Bible, there is a strong link between dominion and service. Jesus came as Lord of all and yet he was also a servant of all.

Humankind is the steward of creation. We shouldn't equate dominion with exploitation. The world is not ours to do what we like with. At the moment, we are raping the world. That's not in God's economy. We need to preserve it for future generations, for them to have a world that's worth living in. If we love God, we will care for his creation – that includes the planet and its poorest inhabitants.

There is a political dimension, but it's not just up to the politicians. We all have to do whatever we can to reduce carbon emissions. We can switch to low-energy light bulbs, drive fuel-efficient cars, use efficient appliances in our homes, not travel more than we need to, use public transport more often, recycle as much as we can, buy our gas and electricity from a green supplier. The world has to get together in a very big way to solve this, otherwise there will be disaster – an increase of floods and droughts and billions of people made homeless.

*With all this talk about wide-scale disaster, where does God fit in? If God is all-loving and all-powerful, why would – why does – he allow millions of people to die through natural disaster?*

We can approach this at a number of levels. We've been put in a world which creates challenges for us. Facing those challenges makes us into the sort of people we are. A person who never experiences pain is a person who hasn't learnt very much. God has also placed us in a world that we can get to know and understand. So we can learn to predict and deal with floods and droughts. We can learn to alleviate the consequences of floods, drought and earthquakes. The fact that we sometimes don't is not God's fault. It is our fault. The fact that we don't radically help those who constantly live with floods or drought is not God's fault. It is our fault.

That, though, doesn't answer personal tragedy, when we call out: 'How on earth could God allow that to happen?' And I ask that question, as much as anyone else. But I have to leave the answer to God. What I do know is that God sent his Son into an imperfect, sinful world to suffer and die in order to redeem the world. The ultimate answer to sin, suffering and pain is the cross of Jesus. I don't understand the cross, but I accept its power.

Human greed is at the heart of global warming. We need to repent of that greed and change our lifestyles. Christians, above everyone, should be doing something about it. Christians could play an enormous part in getting the message of God's love into the world by the way we react to global warming.

*When did you first become enraptured with the world around you?*

I've been interested in science since I was a boy. My first experiment was to make a transformer. I wanted to plug it into the mains but my dad wouldn't let me. So he plugged it in. The result? A loud bang, a cloud of smoke and a damaged chair.

I've always seen science as an outcome of my faith. It struck me that if God was the Creator then science was a study of his creation. I have always tried to put faith and science together: 'How does God work in the world? How does he come into the scientific world? How does he run the operation?' Some people say that science and faith are in conflict but they are not.

*As someone who was 'brought up on the Bible' do you regard the biblical account of creation as being compatible with Darwin's theory of evolution?*

I'm not an expert on evolution but I see no conflict between scientific description and the Bible. There can't be a conflict between genuine scientific investigations and conclusions and what the Bible says, because God is not trying to fool us. The Bible is not a scientific textbook. It wasn't written for scientists as such, it was written for all people for all time.

Some people believe Genesis 1 teaches creation in six periods of twenty-four hours. If it really were that timespan, how is it that on the first day there is an evening and a morning yet the Sun doesn't appear until the fourth day? This suggests Genesis 1 is not a scientific document but a literary – poetic – document. Genesis 1 very clearly describes God as Creator, who makes everything in a very orderly fashion. But it doesn't tell us how or the time-scale in which he did this. The idea that Genesis 1 is to be read literally is very modern. In the third century Augustine had no problem with the idea of creation happening over a much longer time frame.

*Genesis 1 also speaks about humankind being made in God's image. What do you understand by this?*

To me, it means we have some of the characteristics of our Creator. For example, we have a mind and a personality. We can put this the other way round – if we have these attributes, so does God. In *The God Delusion*, Richard Dawkins says he doesn't mind Einstein's idea of there being an intelligence behind the universe. He then says there is no evidence for a per-sonal God, without giving any evidence about what a personal God might mean. But there is all the evidence in the world for a personal God. Believing in an 'intelligence behind the universe' is great but it doesn't get us very far. That is not a God we can get to know.

The wonderful thing about Christianity is that it offers us a God whom we can get to know and who wants to reveal himself to us. That must be the greatest thing in the world that can happen to anybody – to get to know the Creator of the universe.

*You say there is 'all the evidence in the world for a personal God'. Can science prove that there is such a God?*

No, there is no absolute proof of God. If there were, our relationship with him would be different. Science, however, can give perspective. Let's look at the physics of the Big Bang, for example. 'Big Bang' sounds very random and uncontrolled but, in fact, it involved extraordinary precision in order for the universe to be as it is now. If it hadn't been so precisely set up, then we wouldn't have stars and galaxies or Earth, and therefore human life wouldn't have been possible.

The mathematical physicist Roger Penrose calculates that if you work back from the disorder in the universe now – and the fact that the entropy, the measure of disorder, is always increasing – to the beginning, the degree of order had to be set to one part in $10^{(10^{123})}$ [that's 10 times 10 123 times, then 10 times 10 that many times]. That is an extraordinary amount of order and precision. If you started to write a zero on every atom in the universe, you'd run out of atoms. There aren't enough atoms in the universe to describe that number.

For us humans to be here we need the whole universe. The formation of galaxies and stars is responsible for the one hundred-plus chemical elements of the periodic table which constitute life. We couldn't be here if it wasn't for that. This, to me, suggests the existence of a Creator who purposely set things in place so we could exist.

We can also find evidence in the most complex thing in the universe – the human brain. Our brain is such that we can understand something of the laws of mathematics and physics on which the universe is based. Scientists didn't invent those laws; they discovered them. Where did these laws come from? They must have come from somewhere. They existed before humans. They existed before the universe. To me, that's evidence of God as the great Creator and intelligence. Humans also have minds that need to relate to each other and want to relate to a Being who's made it all. The fact that we have minds is also a reason to suppose that there is something beyond what we know scientifically, in merely a material way.

*Some people might think that mentioning God in connection with the Big Bang is veering on the irreverent – a denial of the biblical account of creation. Do you?*

God was the instigator of the Big Bang and has been behind everything that has happened since. God didn't just light a blue touchpaper and retire. He is here, actively sustaining and controlling everything that's going on. Sometimes – through the likes of the miracles of Jesus and the miracle of his resurrection – God intervenes into human history.

*You mention miracles there. What light can science throw on the idea of miracles?*

In general, a miracle is an unusual and significant event that may seem difficult to tie up with the laws of science. Sometimes the miracle is in the timing of an event – something unexpected happens just when we need it. A miracle must have meaning, but usually that meaning is found outside science. Science does not say that things are impossible. If something unusual happens, science might say that there is no explanation for it. Scientists are trained to look for unusual things in case they contain a message.

Science can investigate the historical record of the resurrection, which has been done very thoroughly. But we need to go outside science to explain it. We could use a logical argument for the resurrection. A scientist would recognise the resurrection as an unusual event. But then you ask what it means. If that meaning is very important, relevant and has connections elsewhere then you might well conclude that the event really took place. The resurrection is an unusual event maybe because Jesus was an unusual person. If I then become convinced that Jesus really is the person he said he was, then his rising from the dead is not out of the question.

## FACT FILE

*Professor Sir John Houghton, CBE*
MA (Physics), Oxford.
DPhil in Atmospheric Physics, Oxford.
Lecturer in Atmospheric Physics, Oxford, 1958–1962.
Reader in Atmospheric Physics, Oxford, 1962–1976.
Professor in Atmospheric Physics, Oxford, 1976–1983.
Director General, Meteorological Office, 1983–1991.
Hon Fellow, Jesus College, Oxford.
Fellow of the Royal Society 1972.
Appointed CBE 1983.
Knighted 1991.
Chairman, Royal Commission on Environmental Pollution, 1992–1998.
Chairman/co-chairman, Scientific Assessment for the Intergovernmental Panel on Climate Change, 1998–2002.
President, John Ray Initiative, 2006–present.
Trustee, Shell Foundation 2000–present.
Winner, International Meteorological Organisation Prize, 1998.
Winner, Japan Prize, Foundation of Science and Technology, 2006.
Gold medal, Royal Meteorological Society, 1996.

Gold medal, Royal Astronomical Society, 1995.

Holder of honorary doctorates from twelve UK Universities.

Author of six books including *Global Warming:The Complete Briefing* (Cambridge: Cambridge University Press, 1994) and *The Search for God: Can Science Help?* (Oxford: Lion Hudson, 1995).

12

# Professor Sir Ghillean Prance

## Botanist

He lives on the Jurassic Coast. He camps in the Amazon rainforest. He travels the world to save the planet. He has discovered around three hundred and fifty new plants. Some bear the name *prancei* after him. Former director of Kew Gardens, Fellow of the Royal Society and knight of the realm, Sir Ghillean Prance is the Indiana Jones of botany. Since the 1970s, Sir Ghillean has worked in the Amazon rainforest. He and his family have been welcomed into native communities. He spends a lot of time writing and talking about biblical perspectives on the environment. In December 2007 he attended the UN Climate Change Conference in Bali. Its aims – to get universal agreement on how to reduce greenhouse gases after the Kyoto Protocol expires in 2012 and how to help developing countries in a warming world – are close to his heart.

*Sir Ghillean, how seriously should we be taking climate change?*

When I started as a botanist, the biggest threat was the extinction of species. Even fifteen years ago, that was the case. But today we have the added challenge of climate change. Climate change is driving species into extinction. It is causing sea levels to rise, which threatens the future of whole communities, especially island peoples. If we don't do something about global warming, we will all suffer.

*And what should that 'something' be?*

A great deal of climate change is caused by exorbitant use of energy in Europe, the United States and other developed countries. If we expect the developing world to respect the planet, we have to set the example. If we in the West don't reduce our per capita carbon emissions, then other countries will want to be as polluting as us. Even seemingly small things such as switching off lights, using energy-efficient light bulbs, recycling and using public transport are extremely important. China is developing so quickly that it is building one coal-fired power station a week, which is frightening. But China is also spending a lot at looking for alternatives. It is, for example, using twice as much non-fossil energy as the United States.

*You speak about setting an example. What about your own carbon footprint – flying off to the other side of the world on conferences and field trips?*

That is something that worries me. I try to compensate by making a donation to Climate Stewards, a Christian organisation which plants trees based on the amount of carbon produced in transporting a person round the world. We all need to do more than change the type of light bulbs we use. We've got to change attitudes. There has to be political will for change. I went to the Bali conference to lobby politicians and leaders. Christians should be lobbying more at local, national and international levels. To speak about it more broadly, it's about selfishness. We in the West have got to cut our extravagant, greedy lifestyles.

*The cutting down and burning of the world's rainforests is reckoned to account for some twenty per cent of global CO2 emissions – more than the amount of CO2 emitted by all the world's transport systems. Why is this happening? Is it caused by local people simply making a living?*

The scale is bigger than that. Every year an area of the Amazon rainforest the size of Wales is cut down. Some of it is harvested by local people, but most of the time it is not Brazilians who are cutting down their own forests. Most of the time it is large, foreign corporations.

Currently the big driver of deforestation is the need to clear areas for the planting of soya beans to feed cattle and chickens in Europe and Japan. There is also a huge global demand for timber. In the Amazon, deforestation is driven by big companies who often drive locals off their land. People are being killed over it. The number of small farmers murdered because they resisted is frightening. In 2005, a seventy-three-year-old American-born Brazilian nun, Sister Dorothy Stang, was assassinated because she defended the rights of poor communities against the timber barons. She is just one of many.

*Having spent years living with tribes in the rainforests of Brazil, do you think such communities have a different outlook on the planet?*

I have seen very poor people who are actually much happier – much less worried, less encumbered – than many people in the West. All they want is food and shelter, and they are content. These communities tend to share, while the West is very individualistic. Despite their wealth – and because of greed – many people in the West are living in misery. Becoming wealthier and wealthier does not equal prosperity. The West hasn't taken enough notice of what the Bible says about the dangers of loving money, or of God and Mammon not mixing.

*When did your interest in botany start?*

I wanted to be a botanist when I was a young child. I lived in the countryside and was therefore surrounded by natural history. As children are, I was driven by curiosity. I loved to go birdwatching and to collect plants. As I got to know more about plants, I became excited by the way a whole ecosystem fits together – what pollinates plants, what disperses the fruits and seeds to other places, that sort of thing. It started as a hobby but then I realised I could go on to study it professionally and make my hobby my job, which is fantastic.

*How did your professional life take shape?*

I was director of Kew Gardens for eleven years until 1999. Before that I spent twenty-five years at the New York Botanical Gardens. Currently, I am a visiting professor at Reading University and scientific director of the Eden Project, for which I also run a conservation scheme. That takes me to Argentina for two months of the year, and I spend a similar amount of time working in Hawaii for the National Tropical Botanical Garden. Between fieldwork, I lecture on my work.

*When were the seeds of your Christian faith sown?*

It was while I was studying at Keble College, Oxford. On the first weekend I was invited to the Christian Union. I'd been to church occasionally and read several of C.S. Lewis's books at school but Christianity wasn't a central part of my life. I went along on my first three Sundays at Oxford. On the third one I made a personal commitment to Christ and my life changed completely. In many ways it was completely unexpected. It took me by surprise. I hadn't heard the faith explained in the way that the preacher, John Collins, did. After I made my commitment, I started going to Bible study groups, which made me see how nominal my faith had been as a teenager.

*What does the Bible mean to you?*

The Bible is my final authority. It is God's written word. God has revealed himself in two ways; through his book of words and his book of works. As a scientist, I study his book of works, from which I learn more about how he created things and how the world works. But as a Christian the Bible is my textbook – the guideline – on how I should conduct my life.

*Some people think that believing the Bible to be God's written word is incompatible with believing certain scientific theories, notably evolution. Have you faced such a crisis?*

No, I haven't had a crisis with it, but I have thought about it a lot. Sadly, some of my scientific colleagues are put off the Christian faith because some people say that evolution and the Bible are incompatible; that you can't be a Christian unless you believe that evolution is wrong. But the Bible isn't a scientific textbook. To me, the important thing is that – as the Bible says – in the beginning God created. The Bible tells us that when God created he saw that it was good. Creation is something that brings pleasure to God. How he created is not as important.

*Does the theory of evolution lend itself to a particular Christian or anti-Christian world-view?*

No, of itself evolution doesn't have a spiritual dimension. After Darwin presented it, the Christian author Charles Kingsley said that it was wonderful that God allowed his creation to do its own thing.

*The Genesis account of creation describes humankind as being made 'in the image of God'. What do you take that to mean?*

To me it means that – while there are similarities with, say, the apes – humans are different from the animals. In fact humankind is different from anything else created in two particular respects: we have senses and rationality, and we have a soul.

*What about the other side of that question: What image do you have of God?*

I believe in a triune God. To me, the three-person-in-one God is important because the mission of Christ for a sinful world was vital. I can also say that there have been many times when I have experienced the Holy Spirit. My vision of God is one that embraces all three aspects, and my way to God the Father is through Jesus Christ, God the Son.

*How does your faith impact the way you do science?*

It means I have certain standards of behaviour. It makes one honest in one's science. When I was the director of research at the New York Botanical Garden, I had to make a lot of decisions that had ethical and moral dimensions to them. I once offered my resignation because the garden was fund-raising through the misfortunes of children in Ethiopia. I thought that was immoral. Fortunately my staff backed me one hundred per cent. As a botanist, I am working with God's creation. All around me I see that what humankind is doing to the planet is motivated by greed. I think it is very important that the Church at large, and that I as an individual Christian, should be a strong voice in defence of creation, or 'the environment' as some people call it.

*Let's put that the other way round, how does your science impact your faith?*

The more I learn and delve into the mysteries of how creation works – how things in nature intricately work together – the more it confirms my faith. It is God who brought life into being. Big Bang theory, for example, says that there was a moment of creation. Science is agreeing with the Bible, which teaches that there was a moment when God brought something out of nothing. That should be rather comforting to Christians. What is miraculous is that there are God-given physical laws that hold things together. Those laws make this planet inhabitable, whereas none of the others are. Even though we are discovering more of space, we haven't found anywhere else capable of supporting life. As far as we know, the right conditions for life exist only on Earth. If those conditions changed very slightly, life wouldn't be possible. That, I believe, shows that it is God's doing that we are here.

*To what extent, if any, can science prove that God exists?*

Proving God's existence has nothing to do with science. There are many things – such as love – that we can't prove. But I can say that I have personally experienced God in my life.

*Some areas of science have obvious ethical implications – for example, nuclear energy in physics and human stem cell research in embryology. Does studying the life of plants have moral controversies?*

There are no drastic ones, although at the moment there is a technique that is being abused – the way transgenic plants are made. Manipulating genes – introducing a gene from one plant into another plant because it has a characteristic we like – could do a lot to save millions of people who are starving. But in reality, it is being used to make money for shareholders of multinational corporations. The science is being abused.

*So are you in favour of GM crops, which are so often the subject of Frankensteinian press headlines?*

I am relaxed about the theory of genetically modified crops. I see no ethical dilemma there. But the difficulty is how you use the science. Nuclear power, for example, can make carbon-free energy which could save the world. It can also be used to make weapons which could destroy the world. As far as GM crops are concerned, I am all in favour of making a plant that produces better on arid lands or in salty conditions, or one that is disease-resistant. But the use of GM has been largely motivated by the multinational chemical companies. It is yet another example of greed that dominates the environment.

*Having spent years living in the Amazonian rainforest and more recently making regular trips to the region, how easy is it for you to adjust between the cultures of the West and the developing world?*

I face a real dilemma because one of the difficulties with some of the mission work I see is that Westerners not only preach Christ but they also preach their culture. I believe Christians should go into the world and make disciples but I don't think we should be requiring other cultures to take on Western culture. In fact, some of the ideals these people live by are much more Christian those of the developed world. Most of the tribes I have worked with in Brazil live and work as sharing communities. When a missionary goes in and says, 'You must start a market economy and leave the countryside and move into the town', that is wrong.

I have seen excellent missionaries who have understood the needs and ways of local people. I have also seen mistakes. Back in the 1970s, I ran workshops and in 1993 I published a book called *Missionary Earthkeeping* to encourage people to learn the whole of what the Bible teaches about caring for God's creation – care for the environment and care for individual tribes. Nowadays the approach is better but that's a message the West still needs to hear.

*Christianity is sometimes blamed for the stripping of world resources. This criticism comes because some Christians are said to have taken the account of God giving humankind domination over the Earth as a licence to plunder. How do you read Genesis 1?*

'Dominion' in the biblical sense does not imply domination. The dominion was not God's authority to use up the Earth's resources with no thought for the welfare of other cultures, other creatures, the landscape, the mineral resources, the oceans or the atmosphere. In the Bible 'dominion' means responsible rule, without exploitation. This is reinforced many times throughout the Old and New Testaments. Genesis 2, for example, talks about God making every tree that is 'pleasant to sight and good for food'. Here nature has a twofold function – enjoyment and food – not just a single commercial function.

The story of Noah and the flood, the laws about soil care and farming, the psalmist declaring that, 'the earth is the Lord's and everything in it' and God's reminder to Job of the wide range of the animal kingdom all speak about the wonderful biodiversity that God has made and with which humankind has been entrusted as stewards. The environment was important to Jesus. He knew that nature has much to teach us. For example, when telling his disciples not to go looking for life's meaning or personal fulfilment in things he pointed them to the environment: 'Consider the lilies of the field, how they grow; they neither toil nor spin, yet I tell you, even Solomon in all his glory was not clothed like one of these.' That's a lesson we still need to learn today.

*Environmentalists and green campaigners talk a lot about biodiversity, but what exactly is it?*

Biodiversity is a way of expressing the total variation of life. There are three components. Species is the obvious one and, of course, there are many different species in the world. Then there is variation within a species – genetic diversity. That's why humans look different from each other. Thirdly, there are habitat differences.

The Earth faces a biodiversity crisis. Because we are cutting down and burning the forests that maintain our climate, we are losing the species of plants and animals that will support us in the future. In the face of global warming and environmental disaster, I often ask myself why God doesn't step in and put it right. After all, he stepped into history before and sent his Son when we were in trouble. But it seems that God is entrusting us to put the world right.

## FACT FILE

*Professor Sir Ghillean Prance*
BA (Botany), Oxford.
DPhil in Forest Botany, Oxford.
Knighted 1995.
Fellow of the Royal Society.
Research Assistant, New York Botanical Garden 1963–1966.
Associate Curator, NYBG 1966–1968.
B.A. Krukoff Curator of Amazonian Botany, NYBG 1968–1975.
Director of Botanical Research, NYBG 1975–1977.
Vice-President, NYBG 1977–1981.
Senior Vice-President for Science, NYBG 1981–1988.
Director, the Royal Botanic Gardens, Kew 1988–1999.
McBryde Professor, National Tropical Botanical Garden, Hawaii 2001–2002.
Scientific Director of the Eden Project, Cornwall 1999–present.
Visiting Professor, Reading University 1988–present.
International Cosmos Prize 1993.
Victoria Medal of Honour 1999.
Commander of the Order of the Southern Cross, Brazil 2000.
His exploration of Amazonia include sixteen expeditions on which he collected more than three hundred

and fifty new species of plants. His current research programme is in the Yaboti Biosphere Reserve in Misiones, Argentina.

Author of more than five hundred papers and nineteen books.

# 13

# Professor Bob White

## Geophysicist

He knows the Rock of Ages. He knows the ages of rocks. Seismologist Bob White is Professor of Geophysics at Cambridge University, where for more than thirty years he has been studying volcanoes and earthquakes. His work takes him all over the world. His speciality is studying what happened when continents broke apart to form new oceans, making dramatic natural features such as Fingal's Cave and the Giant's Causeway in the process.

*Bob, what is geophysics?*

Geophysics is the study of how the Earth works. Geology tends to look at what happened in the distant past to create the rocks we see, whereas geophysics looks at what is going on now – for example, earthquakes which are making faults today – from which we can work out what happened in the past. Geophysics also gives us the ability to study things which are too deep to see otherwise. By studying seismic waves, for example, we know what the middle of the Earth is like.

*And how old is Planet Earth?*

The Earth is 4,566 million years old, give or take ten million years. The simplest way of dating the recent past is to count tree rings. We can go back more than eight thousand years by counting them. Counting annual ice layers in the Antarctic takes us back more than seven hundred thousand years. For older rocks we use more than forty different radiometric decay systems. These give us coherent and consistent calculations of the age of ancient rocks and of the planet itself. The oldest life on Earth that we know of is about 3,500 million years old.

*Counting tree rings is one thing. But how do scientists date the universe?*

The technique depends on a mixture of observations and models. One way is to use the microwave background that was created at the Big Bang and observe how it is distributed. From that you can tell how long the universe has been expanding and hence its age. Current calculations suggest that, give or take a few per cent, the universe is about 13,700 million years old.

*Your research focuses on Earth at its most turbulent. Was your discovery of the Creator a dramatic experience?*

Not especially – it was more like a veil being lifted from my eyes so I could see reality clearly. My parents were churchgoers and, growing up, I went to church every week. When I left home for university, I came to Cambridge to read Natural Sciences. It was here, in my first year, that for the first time I met people who talked about having a relationship with Jesus in a real way. At first I thought it was a bit weird to be talking about having a relationship with somebody who was dead and somebody who was God. But I was impressed by the way these same people behaved. They were friendly, supportive and loving towards me. What I didn't realise until later was that they were modelling the love of Jesus. It was the reality of their love which made me think: 'There is something to their lives that is different from mine.'

Then, one Sunday evening I went to hear the Reverend John Stott preach. In a very matter-of-fact way he described what the world is like, explained what God had done in sending Jesus and said that God asked us to respond. And I responded. It was not a big emotional moment; it was more an acknowledgement that Christianity made sense of the world, made sense of my experiences, made sense of my interactions with science and my personal interactions with other people, in a way that nothing else did. And thirty-five-odd years on, the more I walk in the Christian faith, the more sense it makes of the world in which I find myself.

*But are there areas where, like moving tectonic plates, your faith and your science collide?*

No. They don't collide. They are just different ways of looking at God's world. Some people talk about science as explaining how things work and faith as explaining why things work. Science can never tell you why God made the universe. Science and faith are complementary, they are different ways of looking at the same truths. It is, to a large extent, the media that like to show them as being in conflict, mainly I suspect because they think it makes a more sellable story.

*Does the same spark that causes some people to reach out in science also cause people to reach out to God?*

Yes, I believe that is often the case. There are many similarities between the way we do science and the way we develop our faith. Scientists are looking for the truth behind the observations; Christians are looking for truth, for reality, behind their everyday experiences. That's why so many Christians are scientists. I also believe that our ability to understand the world, whether we believe in God or not, is a capability that God has given us – we live in a rational, understandable world.

*One of the subjects you're working on is global climate change. Talk in Britain increasingly centres on carbon emissions, footprints, quotas and credits. Globally, politicians are divided about whether global warming contributes to climate change. How serious is climate change?*

Climate change is something we should take very seriously. The Earth has already heated up by nearly 1°C, and it is headed for more increases. That may not sound like a huge amount, but the average difference between an ice age and the present day is not much more than 5°C. And eleven of the twelve hottest years on record have happened since 1990. The Earth is already warmer than at any time since humans first walked on it. Perhaps an even greater concern is the rate of temperature change. The Earth is getting hotter faster than it has in the past. And it will be the people who are least able to respond who will be worst affected – those who already live marginal lives. Some will die from flooding. Others may starve through drought, or succumb to illness, or be made homeless.

Filling recycling boxes and cutting carbon emissions has a moral dimension. The prosperous high-income regions such as Europe and North America have caused the problem, yet the poorest nations in Asia and Africa will be among those to suffer most. What we do in this country really does affect our neighbour on the other side of the world. Another dimension is who has the moral will to bring about change? Politicians are elected for four or five years at a time. Without wanting to sound cynical, politicians generally do what it takes to get themselves re-elected. They are bound by short-termism and will do only what they think the electorate will allow. Climate change is an issue that requires much longer-term action, with the consequences of what we do now stretching out over many decades into the future.

Britain is responsible for about three per cent of the world's global warming pollution. We should be setting an example and encouraging other countries to change their habits too. It is time for the Christian community as a whole to wake up and get their MPs to push for measures that will slow global warming.

*What is the answer to climate change?*

The distinctly Christian perspective is that there is hope for the future. This is something that is completely lacking from a secular view of the world. In practical matters, we should be looking for ways to do things that don't damage other people – loving our neighbour. Maybe because it produces less carbon, we should look again at nuclear energy. Maybe we should travel less, or use cars less, choose trains over cars or holiday more in the UK instead of abroad. Possibilities go right down to such simple practical measures as using low-energy light bulbs, eating food in season and reusing shopping bags.

*For some people their world ends not because of gradual global warming but through a dramatic event such as an earthquake, volcanic eruption or a tsunami. As someone who studies these things close up, have you worked out why God allows nature to wreak destruction, often killing thousands of people in the process?*

Firstly, nature is not something separate from God. He is sovereign and nature is part of the way he works. Secondly, it is worth remembering that earthquakes and volcanoes are not bad things in themselves. Actually, if it weren't for earthquakes and volcanoes, we wouldn't have all the nutrients that we need to survive. It is because the Earth is mobile and things move around – plate tectonics – that the Earth is not a dead, sterile planet. If it weren't for movements in the Earth's lithosphere – its crust and mantle – this would probably be a lifeless planet. Volcanic regions are often very fertile areas. One of the reasons why people live on volcanoes is because of the fertility of the soil. So they are a positive part of the planet's make-up.

When natural disaster strikes, there is sometimes an element of human sinfulness that magnifies the suffering. In 1999, the Izmir earthquake in Turkey killed more than seventeen thousand people. Some of the photos at the time showed three hundred-year-old minarets standing while surrounding modern tower blocks had collapsed, killing hundreds of people. An investigation later found that, although building codes to protect against earthquakes were in place, some builders had cut corners by bulking out the walls with old oil cans to save on concrete. Had those buildings been built to standard, many fewer people would have died. But somebody got greedy. In the 2004 tsunami, a lot of deaths were due to the fact that people had cut down mangrove swamps and were living in places they shouldn't have been. So it is a more nuanced subject than people sometimes give allowance for.

*Nevertheless, why does God allow suffering?*

Part of the answer lies in the fact that God did not create humankind as puppets. He could have but he didn't. He chose to make the world one where we have the ability to respond to him. But humankind as a whole has chosen not to respond to God – what the Bible describes as the fall. One of the consequences of that is that the whole of creation – including the natural world – is somehow out of kilter with God because of humankind's sinfulness.

God made a good world but we are not free of the consequences of our actions. God also offers hope in the form of a new Heaven and a new Earth where, as the Bible puts it, for those who have put their trust in him, there will be no more crying and no more pain.

*Some people look to the Bible to provide a timetable as to when the world is going to end. One of the signs they suggest is an increase in earthquakes. Is the frequency of earthquakes on the increase?*

No. It's just that we have better instruments now so we can detect them better and can detect smaller earthquakes so we record more of them. It's possible to plot the size of earthquakes against the frequency with which they occur. An earthquake with a magnitude above eight happens on average about once a year. I investigate very small earthquakes – magnitude two and less. These are hundreds of thousands of times smaller and they happen millions of times every year. In the distant past we know there have been earthquakes above magnitude nine, like the one that caused the Boxing Day 2004 tsunami. There is no evidence that the rate of occurrence of earthquakes has changed. Using the Bible to work out when the world will end is an abuse. Jesus said he didn't know when the end would come, so why should we expect to know? He did say, however, that we should live as though it was going to happen tomorrow. Which seems good advice to me.

Global warming will not bring about the end of the world, but it would mark the end of many communities. The planet will survive, but its poorest inhabitants may not. One of the striking things about the Earth is that throughout the past four thousand million years it has maintained an average temperature between 0°C when water freezes and 100°C when it evaporates. The sun has got about thirty per cent hotter during that time but the Earth has maintained that narrow temperature band, which makes it possible for life to exist here. That is amazing. Indeed, over the past eighty million years the average temperature has remained within an even narrower range, varying by less than 10°C from the present temperature, making it possible for mammals to flourish.

The universe is extremely finely balanced. The anthropic principle, which is a subject of scientific study, is concerned with some of these incredibly precisely tuned parameters that make life possible. Someone once suggested that the tuning of some of the physical constants is as precise as the accuracy required to hit a one centimetre square target with an arrow fired from the other side of the universe. There has to be the right force of gravity, the right forces of attraction and repulsion in the nucleus, the right electrical forces and so on, otherwise matter as we know it, and ultimately life itself could not exist. One way of looking at that is to say that we're here just by chance. But as a Christian I see it as evidence that a gracious and loving God created a place just right for us to inhabit.

*Would you use such an insight as proof of the existence of God?*

No, not really, though, of course, it is completely consistant with a loving Creater-God. If there were a knock-down proof for God, presumably we would all be Christians. The theological reason why we're not is that our eyes are veiled so we can't see the spiritual reality that is in front of us. We depend on God to remove that veil and enable us to see the reality. That is what God does when people become Christians. While I can tell people about God's goodness – tell them the good news of Jesus – I cannot myself convince them of the reality of God. Only God can do that. I can't prove that God exists, but I can say that everything I know about this physical world, my own nature and my interactions with other people is consistent with God's existence and with what the Bible tells us about God's purposes for us and for his creation. To me Jesus, and only Jesus, makes sense of this life.

*Science offers us a number of theories or models about how the world came into being. Do you believe that God created the universe?*

Yes. My faith depends on knowing that God is sovereign, that he made this universe and that he will remake it in the fullness of time, as a new Heaven and a new Earth. It's important that God created it, because the alternative is that life is random and meaningless. To live with that kind of hopelessness would be dreadful.

*A 'new Heaven and a new Earth' is a phrase straight from the Bible. How do read it – as a metaphor or will it happen physically?*

I think it is real. The Christian view is an holistic one – you can't be a person without a body and a soul. We will not be floating about in Heaven as disembodied spirits. Obviously, when the Bible talks about a new Heaven and a new Earth, it is describing things that are beyond our experience. But it uses very down-to-earth terms. It says that there will be meadows, trees, animals and people. Interestingly there will also be a city. Cities are places of human creation. The Bible even gives a sense of continuity between this world and the next when it says that the honour and glory of the nations will be taken into the new creation. It means that what we do now, in this world, has eternal significance.

I'm looking forward to watching volcanoes and earthquakes that won't kill anybody. What a marvellous sight they will be! I imagine the physical world of the new creation will be rather different from this one. After all, the only person who has come back to this world from the next – Jesus himself – could walk through walls. But he also did normal things – he ate meals and chatted with friends. Jesus also bore the scars of his crucifixion, so something that had happened to him on Earth was carried through after his resurrection.

*Another biblical image is of humankind being made 'in the image of God'. What do you understand by it?*

To me, it means we have some of those attributes that God himself has, such as the ability to love and be loved, great creativity and a concern for justice, which is one of the big things that differentiates humans from animals. These aspects mean that we can relate to God in a personal way. It also means that we should be stewards of the Earth, ruling over it on his behalf, as his first command to humans made clear in Genesis 1:28.

*One of the theories that tries to explain how life works is evolution. Darwin gets a bad press in some sections of the Church. How do you view evolution? Is it necessarily atheistic?*

Richard Dawkins thinks it is, but I think he's wrong. Part of the reason some people don't like evolution is the idea that it is completely random. But if as a Christian you have a strong view of God's providence, then 'random' needs redefining. Things happen in my life, for example, which I see as providential but which someone else might say is mere coincidence. If I have a strong view of God's sovereignty, then who am I to say he could not choose evolution to create all living things?

*Did humankind evolve from something in the primordial slime or were we created by a special act?*

I believe in an historic Adam and Eve. I believe in the fall. I also believe that evolutionary scientific theory is a very powerful way of explaining the relationships between all living organisms. The theory of evolution has been around for a long time. We keep getting more tests on it and the theory gets stronger each time. For example, we can test the human genome (something Darwin had no conception of being able to do) and discover that it is 98.5 per cent the same as that of chimpanzees in terms of DNA. We also have retroviral inserts in our genomic structure that can be tracked back through our ancestral species – a sort of living historical record of our evolutionary past. But I believe passionately that humans are more than just animals. We are animals plus we're made in the image of God.

I believe in God's sovereign activity in the world but I don't think God parachuted on to the planet two ready-made people who had DNA very close to the animals. That would have been trickery. My view is that he breathed life into animals and turned them into humans – made them have the ability to respond and relate to him, made them morally responsible for their actions – and human his-tory goes on from that moment.

*How do you view the creation accounts in Genesis?*

Genesis was not written as a scientific account of the world coming into being. Rather the abiding message of Genesis is a theological one: that God is Creator and that he purposely made the universe of his own free will. It is written as a narrative of a workman who does a hard day's work, sits down at the end of each day and says: 'That was good.' When he finishes by creating humans, he says: 'That was very good,' and has a rest.

Genesis 1 is full of arguments against the prevailing views and myths of how the world came about. It says that the world is not an accidental place, that humans weren't just created as slaves or playthings by the gods, that the universe has meaning and purpose behind it. There are also many little clues that it is foremost a great theological statement. For instance, when Genesis 1 was written there were perfectly good words around for 'Sun' and 'Moon'. But at the time people worshipped the Sun and the Moon as gods. So when the writer of Genesis described the Sun and the Moon he says that God created the greater light to rule the day and the lesser light to rule the night, making it clear that these were the functions of the Sun and Moon and that God didn't create other gods.

Another hint about the theological purposes of the creation account is that in those days people were scared of the sea. But Genesis says that God created the sea and all its creatures – including the big scary ones like leviathon – so there is reassurance that he is in charge of them too. Genesis 1 answers some contemporary myths. A New Age attitude is to worship the creation, but Genesis points to the Creator as the rightful object of worship. Post-modernism says that meaning in life comes only by what you make of it. Genesis 1 says that God created the world for a purpose. He meant to create it and it was good.

*And did God do all this within six 24-hour days?*

The six days are a literary device of a week's hard work. In any case, as a geologist I find it amusing to reflect that when the Earth was created it was spinning much faster than it does now, so one day lasted only about five hours. Due to the gravitational pull of the moon and the tides, which absorb energy, the Earth's rotation has been slowing down ever since, so actually six 24-hour periods wouldn't fit either!

*Having travelled the world with your work, what insight does science give you into miracles?*

It doesn't surprise me that God uses his creation to work his miracles. The Bible actually explains how God did one of the great miracles, the Israelites crossing of the Red Sea. It says he made a strong east wind to blow and make the waters move back. God is the Creator of this universe, so you would expect him to use his creation to work his purposes. When the Israelites crossed the Jordan, the river dried up and they walked across on dry land. The Bible says that the river was blocked further upstream. As a scientist, I know that earthquakes occur in that region that cause landfalls that dam the river often for several days upstream from where the

Israelites crossed. To me, such an explanation does not take away the wonder of God's timing and intervention.

*What is your working definition of a miracle?*

A miracle is something outside normal scientific happening. There are miracles that we can't explain, supremely the resurrection of Jesus. The resurrection is pivotal to Christian belief. It is God intervening in human history and saying: 'I have a plan for the future of this Earth – a new creation where everything will be put right.'

*Bob, how does what you learn in science affect your life as a Christian?*

As a geophysicist I spend my life talking about things that are millions and billions of years old, yet humans have been around only for the last little fraction of that time. This makes me realise that God not only loves and takes pleasure in all of his creation, including the non-human and non-living part, but also considers humankind as very special – that the long evolutionary history of the world was moving it forward to be a place where humans could flourish. The psalmist talks about God flinging stars into space. Science tells us that we need the billions upon billions of other stars in the billions of other galaxies so that conditions for life here on Earth are just right. Science also tells us that our bodies contain carbon and other atoms that were created in stars billions of years ago. All that time was needed to get things just right for humans to live here on Earth. That is awe-inspiring. Things like this make me realise how fruitful and generous a creator our God is.

## FACT FILE

*Professor Robert (Bob) White*
BA (Natural Sciences) Cambridge 1974.
PhD (Geophysics) Cambridge 1977.
Fellow of the Royal Society.
Fellow of the Geological Society.
Professor of Geophysics in the Department of Earth Sciences, Cambridge 1989–present.
Associate Director of the Faraday Institute for Science and Religion, St Edmund's College, Cambridge.
Director of the John Ray Initiative.
Leader of research group investigating the Earth's dynamic crust: in particular the earthquakes that result from Earth movements and the way in which enormous volumes of volcanic rock are produced when continents and oceans rift apart. His fieldwork at sea has taken him to the Atlantic, Indian and Pacific Oceans.
Currently investigating the internal structure of volcanoes in Iceland, New Zealand, and the Faroe Islands.
Author of more than three hundred scientific papers and articles. Co-author with Denis Alexander of *Beyond Belief: Science, Faith and Ethical Challenges* (Oxford: Lion Hudson, 2004), and with Nick Spencer of *Christianity, Climate Change and Sustainable Living* (London: SPCK, 2007).

# 14

# Professor Colin Humphreys, CBE

## Materials scientist

To enter Colin Humphreys' study is to walk into an Aladdin's cave. It is crammed with scientific toys, models and gizmos. Some are leaving gifts from his former students, others are of his own making. Colin, who is the Goldsmiths' Professor of Material Science at Cambridge University, travels the world to share his knowledge. Aware of the need to cut globe-trotting carbon footprints, he is working on aircraft engine alloys with Rolls-Royce. The plan is to find materials to develop fuel-efficient, high-temperature aero engines. From a carrier bag he pulls his latest project, a credit-card-sized torch with an amazingly bright white light. It's the sort of thing adventurous boys and girls the world over would like for Christmas.

*Colin, what's so special about this torch?*

We are trying to replace all the lighting in the world with this. My main area of research is on gallium nitride, which emits brilliant lights of different colours. Gallium nitride is not a natural compound. We have to make it, which is quite a challenge. White light-emitting diodes (LEDs) are already in use in the flashes of mobile camera phones and bicycle lights. I am working on higher-quality white LEDs which give off a less harsh light by mimicking natural daylight. The hope is that they will replace the standard light bulb, even the energy-efficient ones.

Lighting consumes twenty per cent of all electricity output. With LEDs we could halve that figure. Electricity prices, CO2 emissions and the replacement of power stations could be cut by ten per cent. That's a bigger reduction than would be saved through wind power. Because higher-quality white LEDs mimic daylight, people who suffer from lack of sunshine in the winter – seasonal affective disorder – would be healthier. The bulbs last up to sixty years. Low-cost, long-life, high-quality, environmentally friendly lighting will mean that these LEDs are the ultimate light source.

*That sounds a little bit like* Tomorrow's World *– where some supposedly life-changing inventions were showcased on TV, never to see the light of day again. What are the chances that your gallium nitride light will catch on?*

Oh, it will happen. Semiconductor companies are behind this and light-bulb manufacturers are following. When Philips, for example, produced LED flash units for camera phones, sales rose to one hundred million units in less than three years. Audi are using LEDs in daytime headlights. BMW and Lexus are to follow suit with main headlights. LED lighting is used on yachts. In 2006, the front of Buckingham Palace was illuminated with LED lighting for less than the cost of boiling a kettle. Getting this technology into people's homes depends on reducing production costs, but it will happen.

*Besides being used in what we might call luxury goods, what other implications does your LED lighting have?*

It could radically change life in the developing world. The average person in India has less than three per cent of the lighting of the average person in America. To get even twenty per cent in conventional ways would mean building six times the number of power stations. But LED lighting doesn't need connection to a national grid. It can run off a solar panel. So people who can't work once the Sun goes down will be able to have light and it will cost nothing to run.

We've also discovered that at deep ultraviolet levels gallium nitride kills all viruses and bacteria. In the developing world impure water is a bigger killer than Aids. One idea would be to have an array of ultraviolet LEDs on the inside of waterpipes. The water would still look filthy but it would be bug-free and safe to drink. Another application would be to kill hospital superbugs or to purify air. It is difficult work and some way off yet, but it will be possible.

*Colin, when did you become a Christian?*

My parents were Christians and as a lad I went to church and Sunday school. When I was about fourteen, I decided to be a Christian. But when I went to study at Imperial College, London I threw it overboard. For about a year I said, 'I am not a Christian: it is all fairy stories.'

One night one of my flatmates asked me to go to church with him. I knew that some good-looking girls went to this church so I agreed. After the service a group of us went back to the minister's house. A week or two later I went again and was walking home from the minister's house with one of my colleagues when he said that he wanted to be a Christian and I should tell him how to go about it. I thought to myself: 'What do I say? He thinks I am a Christian. I know I'm not. Do I tell him I'm not a Christian and spoil his chance of faith? Or do I tell him what to pray but not have any conviction in it myself?' In the end I told him to speak with the minister. But it got me thinking.

I sat up all night. His question challenged me. Was I a Christian or not? That night I decided I would become a Christian, and I started reading the Gospels and Christian books. I confirmed this decision later. I later learnt that my friend had seen the minister and become a Christian.

*How does your faith affect your science?*

I hope it affects the way I deal with people and the conflicts which can arise with people when you are a scientist. I am also delighted that I am working on something which will directly benefit people. For many years that wasn't the case, and it troubled me.

*And how has science had an impact on your faith?*

I think scientists see the world slightly differently. To me, Christianity is a very logical and reasonable faith. When I read the Bible, I tend to look for natural explanations. That doesn't mean I don't believe in miracles but that I think God often works in and through nature to achieve his purpose. Often the miracle is in the timing of an event. And there are some things, like the resurrection of Jesus or the virgin birth, for which we can't give a scientific explanation and which certainly are miracles.

*What is your working definition of 'miracle'?*

I go with Aristotle. He spoke about prime movers and agents. He said that God is the prime mover and agents can be things like natural events. So how do you know an event is a miracle or not? It is determined, said Aristotle, by the timing. If, for example, you look at the story of the Israelites crossing the Red Sea, the Bible says that Moses stretched out his hand over the waters all that night. God sent a strong east wind to drive the waters back. So it was a natural event – there's no hint of God sending an angel or clicking his fingers. God was using nature to accomplish his purpose. But that it was done at the precise moment when the Israelites were surrounded and about to be crushed by the Egyptians makes it a miracle.

There's another principle at work, which we can see if we look at the time Jesus walked on water. Often people describe this as a miracle by saying that God stopped gravity operating, otherwise Jesus would have sunk. I look at it another way. Rather than the natural phenomenon being suspended, why not say God provided an additional force which upheld Jesus? That way God doesn't break his own rules to perform a miracle.

*So, for you, is God someone who is more likely to use what he has already created rather than someone who is forever intervening to get people out of trouble?*

It is a difficult subject. If prayer is meaningful – if prayer can change things – then there has to be flexibility in the system. I think that some things are fixed. God plans things. I believe, for example, that God planned that Jesus would come into history at a certain point of time and would be crucified when he was. I also think that God leaves a number of things flexible, that he interacts with us to achieve his purposes. For example, like everyone else I've made good decisions and wrong decisions. If I had always made the right decisions, my life would have followed a certain course, which may well have been a better course. But there again I believe God works with us, if we let him, to make the best of a wrong decision. Because God has given us free will, life cannot all be rigidly planned. For example, the Bible says that God wills for the whole world to be saved. But it hasn't happened yet because humans exercise their free will and reject God. And God allows that.

*Do you apply this mix of direct divine intervention and God allowing 'natural' processes to continue to the way life has come about?*

As a Christian, I believe God is in charge. He worked with his created order for the emergence of life, including humans. He planned from the beginning that humans would emerge at a time and in a way of his choosing. I was brought up to believe that the Earth was around six thousand years old. It was one of the reasons why, when I went to university, I thought the faith wasn't true. I have a great deal of sympathy with creationists because I know they are very genuine people. But I now believe that evolution is the way God developed life.

*Why do some people see evolution as an attack on their Christian faith?*

If someone interprets the early chapters of Genesis and concludes that the Earth was created around 4,000 BC and evolution requires longer for life to develop, then both accounts cannot be right. Evolution does not say that humans evolved from apes. It says that humans and apes share a common origin. For non-believers, evolution is taken as evidence that life is nothing more than a series of blind chances and is, therefore, meaningless. From a faith perspective, I see evolution consisting of a lot of events which look like chance but which are part of a process guided by God.

*The creation accounts in Genesis include the idea that humans are made in the image of God. What do you understand by this?*

I am tempted by the way the ancient Egyptians used the same phrase. The Egyptians had many gods. They made statues and images of their gods, and worshipped them. To them, these images were representatives of God on earth. I like the idea that in the Genesis account, humankind was made to be the representative of God on earth.

*Sometimes the public gets worried that one group wants to be more than God's representatives, when scientists play God. Are you worried that scientists can go too far?*

Scientists will increasingly have the ability to play God, especially in the biological realm. We will be able to do remarkable things which can be used for good or evil. It is vital that there is moral guidance. There is also a great need for Christians to be at the forefront of science in the future.

*In your work you're trying to solve the problems of the future. But in your spare time you've also been researching events of past, particularly the star of Bethlehem. What have you discovered? Is the biblical account surrounding the birth of Jesus true or is it a story that tells a truth?*

I take it as being literally true. In Matthew's Gospel, the account is sandwiched between a genealogy of Jesus in chapter one and the coming of John the Baptist in chapter three. Both of these are rooting Jesus in history. Chapter two opens with, 'After Jesus was born in Bethlehem in Judaea, during the time of King Herod . . .' King Herod reigned from 40–4 BC. There was another Bethlehem but Matthew specifically identifies 'Bethlehem in Judaea'. The account has all the hallmarks of a story that is meant to be taken literally. Jesus is being placed in history and geography – real time, real places, real people.

There are three possible explanations for the star. One, it was a mythical star – an invention that didn't really exist. Two, it was a miraculous star – one which God created just for that occasion. Three, it was a real astronomical object. I firmly believe that God could have produced a miraculous star, but with ancient Egyptian and Babylonian literature containing stories of new and special stars. As far back as the 1600s Austrian mathematician and astronomer Johannes Kepler started applying his newly developed laws of planetary motion to identify the star of Bethlehem. My approach was: Is there a known star which fits as precisely as possible the description we have in Matthew's Gospel?

*But how much workable scientific data does Matthew's account give us?*

A lot of commonly recited details about the story don't come from the Bible at all but from Christmas carols. The Bible doesn't say that the star guided the men. That's from 'As With Gladness' and 'We Three Kings'. This raises the question of the star-followers. Were they 'wise men'? Were they kings? The 'Magi' – as the Gospel calls them – were almost certainly Babylonian astrologers-cum-astronomers. Unlike today, back then people didn't distinguish between astrology and astronomy.

In Matthew, the star has three distinctive, unusual characteristics. Firstly, Herod asked the Magi the exact time the star had appeared. This strongly suggests it was a newly appeared star. Secondly, it appeared to have travelled in the sky because the Magi first saw the star in the east and then went to Jerusalem. Then from Jerusalem they travelled due south to Bethlehem, where they saw the star ahead of them. Thirdly, as they went towards Bethlehem, it is said the star stood over the place where the child was.

If it was a new star, there are a limited number of options it could be: a nova, a supernova, a meteor, a shooting star or a comet. If the star travelled in the sky, it couldn't have been a nova or supernova, which explodes but does not trace across the sky. Comets, on the other hand, whatever their orbit, travel one or two degrees a day through the sky against the star background.

The clue is the phrase, 'the star stood over the place where the child was'. Many Christmas card designers have interpreted 'the place where the child was' as the stable, but my understanding of the original Greek is that it could equally be taken to mean the town of Bethlehem. Either way it works, as the Magi travelled from Jerusalem to Bethlehem.

*How common a phenomenon, though, is it for a comet to appear to stand over a certain place?*

I found only two references in ancient literature to stars standing over places. One was by the Roman historian Cassius Dio, who said a star stood over Rome. This is regarded as a reference to Halley's Comet and dates the occasion as 11 or 12 BC. The other comes from the Jewish historian Josephus, who talks about an event which occurred in AD 64. He said a star like a sword stood over Jerusalem. We know from other ancient literature that comets were frequently described as swords because they can have a long pointing tail. Historians agree that Josephus was describing a comet. In fact, through Chinese astronomical records, we can identify this particular comet. So, on the two occasions when a star is said to have stood over a place, it was a comet. It is very likely, therefore, that the star of Bethlehem was a comet and that its tail pointed out Bethlehem.

*What does identifying the star as verifiably real prove?*

It is an indication that Jesus really lived. It helps place him in history. It means we can get an idea of when he was born. Chinese astronomers always recorded whether a comet had a short or a long tail. To the Chinese, the length of the tail and its position in the sky were significant. A long-tailed comet was known as a broom star. To them, it signified the sweeping out of the old and the start of a new era. Between 20 BC and AD 10 they recorded three comets. In this period there was just one long-tailed comet. It appeared in 5 BC and was visible for seventy days. King Herod died in 4 BC, so the star from the Chinese record fits the time frame well.

*Speaking of time frames (and acknowledging the AD/BC contrivance by which the birth of Christ is usually dated around 4 BC) how does the appearance of the 5 BC comet fit with the biblical account that Herod, wanting to wipe out the threat of a rival king, ordered the killing of all boys aged two and under?*

This is Herod taking no chances. The question is: Why did the Magi come to Jerusalem to ask about a new king in the first place? I think this is down to a sequence of three events. In 7 BC Saturn and Jupiter lined up three times: astronomers call this a triple conjunction of planets. (Some people consider this as the star of Bethlehem.) In that year, Jupiter and Saturn came very close together three times. To astronomers-cum-astrologers at the time, that would have been very significant. Babylonians worshipped Saturn as the main god and Jupiter as his son. The symbolism of Saturn and Jupiter lining up closely together and then separating was that a son of God was to be born. For this to happen three times was a sign of the importance of this birth.

Secondly, the Babylonians associated the constellation of Pisces with Israel. When the planets lined up, they did so with Pisces in the background. This was taken as a message that the son of God would be born in Israel. Some years ago, archaeologists found a clay tablet called the Star Almanac of Sippar, at a place called Sippar about thirty miles from Babylon. It records the conjunction of Saturn and Jupiter as happening in 7 BC. The clay tablet is now in the British Museum. Whether it was a prophecy given in advance or something written after the event, we don't know; but it is significant that people considered the lining up of Saturn and Jupiter in Pisces important enough to write down. In 6 BC, Mars joined Saturn and Jupiter in the sky, and all the planets were very close together. This is called a planetary massing. Mars was the sign of a mighty warrior, so the message was that this would be a mighty king.

The question for the Magi was: When is this going to happen? This is answered when the comet appears in

5 BC, because a comet in the east meant to the Babylonians that the event will happen now. They then make their way to Jerusalem and tell Herod, 'We have seen his star in the east'. Matthew's Gospel records that Herod questioned them about the 'exact time' the star had appeared. The fact that Herod ordered the death of all boys aged two and under suggests that he – along with the Magi – considered the 7 BC conjunction of the planets as a sign of the birth of a new king.

*When the 'wise men' saw the star they recognised its significance, they followed it and they discovered Jesus. Why is the star of Bethlehem so special to you?*

To me, it speaks of how God uses natural phenomena. He could have made a special – miraculous – star but he didn't. He used what he'd already created: planets and stars. Significantly, God chose to use Gentiles – foreigners – to herald the birth of Jesus. This speaks of God being inclusive. He sent Jesus for the whole world and whole-world citizens are represented in the events of his birth. Being able to identify and date the star adds scientific weight to the fact that Jesus was born where and when the Bible says.

## FACT FILE

*Professor Colin Humphreys, CBE*
BSc (Physics), Imperial College London.
PhD in Physics, Cambridge.
MA, (Metallurgy), Oxford.
Appointed CBE 2003.
Fellow of the Royal Academy of Engineering 1996.
Goldsmiths' Professor of Materials Science in the University of Cambridge 1991–present.

Chairman of the International Advisory Board of the National Institute for Material Science (NIMS), Japan 2003–present.

Member of the International Advisory Board of the Institute for Materials Research and Engineering (IMRE), Singapore 2004–present.

Director of the Cambridge Centre for Gallium Nitride 2000–present.

Director of the Rolls-Royce University Technology Centre for Aerospace Materials 1994–present.

Author of more than 550 research papers, and the book *The Miracles of Exodus: A Scientist's Discovery of the Extraordinary Natural Causes of the Biblical Stories* (London and New York: Continuum, 2003).

# 15

# The Reverend Dr Rodney Holder

## Astrophysicist

For years he has watched *The Sky at Night* and studied the night sky. Now he's encouraging others to do so. For fourteen years, he worked as an operational research scientist for the Ministry of Defence. Then God called him to be an Anglican priest. After ordination he took a number of pastoral appointments. Since 2006, cosmologist the Reverend Dr Rodney Holder has been the course director of the Faraday Institute for Science and Religion, based at St Edmund's College, Cambridge. The college was also the home during his time in Cambridge of another scientist-priest, Georges-Henri Lemaître, whose work led to the Big Bang theory.

*Rodney, when did you first become fascinated with science?*

I was interested in astronomy from the age of about seven. I had a small telescope and read all the Patrick Moore books. At university, I read maths at Cambridge and went on to specialise in astrophysics and cosmology at Oxford in a theoretical astrophysics research group led by Dennis Sciama, who was an inspirational supervisor.

*As well as researching the interaction between cosmology and theology, you are providing opportunities for people of faith to engage with science, and for scientists to engage with theological ideas. Presumably, you don't see faith and science as two conflicting worlds.*

No, science and faith interact at many levels. The common perception that there is a conflict between faith and science is fostered by two kinds of people. One kind is the militant atheist – the Richard Dawkinses and Peter Atkinses of the world – and the other kind is the Christian who claims to read the Bible literally. I think both are mistaken. The atheists say that science can answer all the kinds of questions we want to ask. This makes religion redundant. But that statement is clearly false. Science can't answer ethical or spiritual questions. Then there are those who say that if we take the early chapters of Genesis literally, then the Earth is only a few thousand years old and evolution is wrong. That kind of approach is a complete misunderstanding of the Bible.

As an Anglican priest, I believe the early chapters of Genesis are the word of God. The issue is how we interpret those chapters. Genesis was written to be understood by the people at the time. To get to a true interpretation, we need to understand the surrounding mythologies which the Israelites encountered. We need to know what Genesis was attacking in terms of polytheism, the idea that matter was evil, stories of gods battling each other and people worshipping the Sun and the Moon as gods.

In the Genesis account of creation, for example, the Sun and the Moon are not named but are called lights. If they were named, they would be the names of gods which were there to be worshipped. Today people read horoscopes, they believe their lives are governed by the stars. The message of Genesis 1 is that the Sun and Moon are not gods that govern our lives but are simply lights.

In non-biblical ancient stories you find that human beings are made to serve the gods, to provide them with their food. In Genesis it is the opposite: human beings are given supreme responsibility to exercise dominion – stewardship – on behalf of God over what he has made.

*Do you read the Bible the same way you read a scientific textbook?*

No, the fundamental message of Genesis is theological. It is not telling us a scientific story but conveying a vital theological message – one that is still true. There is a repetitive strain in Genesis 1: 'God saw that it was good.' Genesis tells us that it is God who created; that (in contrast to the pagan myths) the creation is good; and that human beings are made in the image of God. Genesis also tells us that the relationship between human beings and God has been soured by our rebellion – the fall. This is a vital foundation for what follows in the rest of Scripture.

*You speak about humans being made in the image of God, but what is your image of God?*

For me, the ultimate way to see God is to look at Christ, where we see God revealed as a Person, as one of us. The most staggering thing of all is that the God who made the universe in all its vastness also became human in Christ and came to redeem us from the effects of the fall.

*What about your own redemption; how did you come to faith?*

It was a sudden conversion, which came through a sermon I heard shortly after I went to Cambridge as a student. When I arrived, I was sceptical about faith. But I was also idealistic and I was looking for something with real meaning and purpose. During the first week or two of term, I went to a freshers' service and heard the gospel preached by a vicar called John Collins. The message of salvation in Christ – how his death reversed humankind's rebellion against God, and how out of love he had died so that I could be forgiven for my wrong-doing – was incredible. It really hit me. I responded and invited Christ into my life. That was just the beginning, of course. Although it seemed to have come out of the blue I've always wondered whether in fact some of my Christian friends had been quietly praying for me to come to faith.

*What were the circumstances of God calling you to the ministry?*

Even though I enjoyed work as a scientist, there'd always been a nagging thought about doing more for God. Years ago, I'd had discussions with a college chaplain. I was always trying to grapple intellectually with the faith, so I was never averse to considering science and faith in each other's light. Through studying astrophysics, I saw the grandeur of God in creation – the vastness and scale of the universe. For me, science just enhanced what we know about God and his majesty. Then in 1993, I was writing a book and doing a lot of theological reading. Things at work were changing in ways I didn't like, and I began wondering whether God was saying something to me through all of this.

Another influence was the fact that my wife, Shirley, suffered an ectopic pregnancy and nearly died. We were also told we wouldn't be able to have children. That put a whole new perspective on life. We asked ourselves what the absence of children could mean in terms of our commitment to God, and we thought it made us freer and more flexible. (Ironically, when I did eventually get to theological college, I was surrounded by couples who were producing children at a phenomenal rate!) Shirley and I spent six months praying about whether I should offer for ordination. I eventually trained at Wycliffe Hall in Oxford.

After ordination, I spent four years as a curate in Warwickshire and seven months as chaplain to the English Church in Heidelberg, Germany, before, in 2002, becoming priest-in-charge of the Claydons, a group of villages in Buckinghamshire.

*What does your faith give your science?*

Faith gives me the motivation to study science. Over the years, religious faith has been a motivation for many people to explore science because this is God's universe. If it isn't God's creation, why aren't we living in a chaos rather than a beautifully ordered and structured universe, whose laws we can uncover? Those laws for me show God's faithfulness to us.

*And what have years of deep-space science done for your faith?*

Scientific discoveries, such as the fine-tuning of the universe, feed my faith and deepen my appreciation of God. For example, the universe in its initial phase – the first fraction of a second from the Big Bang – had to be set up in a very special way in order for stars, galaxies and ultimately life to form. Cosmologist Fred Hoyle did some major work on the nuclear reactions that go on inside stars to form all the chemical elements out of the simplest building block, which is hydrogen. He discovered that there needs to be a very fine balance of the forces in nature in order to make carbon, and then to make oxygen without destroying the carbon. Although he didn't believe in God, Hoyle said that his work led him to the conclusion that there was a super-intellect behind physics, chemistry and biology, and that there are no blind forces worth speaking about in nature.

*To what extent can science prove there is a God?*

As a Christian I can't prove the existence of God, but as a scientist I see the fine-tuning of the universe as a very powerful suggestion that there is a God. The universe is sitting on a knife-edge as to whether it is going to expand for ever or eventually re-collapse. This is to do with how much total mass-energy there is in the universe. If there is more than a certain critical value, then gravity will pull the universe back and it will re-collapse. If there is less than that critical value, then gravity won't be sufficient to pull the universe back, and it will expand for ever.

Right back at the beginning, the universe needed to be very close to that knife-edge in order for stars, galaxies and planets to form. The mass-energy needed to be what it was to within 1 part in $10^{60}$ (that's ten with sixty noughts after it). That kind of accuracy would be the same as firing a gun from one end of the universe to the other (some ten billion light years away) and hitting a coin you were aiming at. The question is: Was that a lucky shot or are you a brilliant marksman? Likewise, is the fine balance of the universe a lucky happenstance or is there a brilliant designer behind it? I believe that there's a brilliant designer behind it.

*Cosmologists calculate that there are one hundred thousand million stars in our galaxy and one hundred thousand million galaxies in the observable universe, all to sustain life on Earth. Why does this planet need so many other heavenly bodies?*

You might think that a universe with one galaxy and one hundred thousand millions stars would be enough. But there is a simple relationship between the mass, size and age of a universe. A one-galaxy universe would be only about a month old. You need the universe to be as old as it is – 13.7 billion years old – and as vast as it is to have life anywhere. In a way this reverses Copernicus's discovery that the Earth is not at the centre of the universe. We know now that the Earth revolves round an ordinary star somewhere near the outer edge of an ordinary galaxy. It's not special by being at the centre.

But life on Earth – life anywhere – couldn't exist without all those one hundred thousand million other galaxies. It's as though the universe revolves around supporting life on Earth (and possibly elsewhere).

*Some scientists say that our universe isn't that special because it is one of a number of universes – the theory of the multiverse. Would you explain that, please?*

The multiverse theory says that a whole ensemble of universes might exist among which the initial conditions and ratios of the forces of nature vary from one member universe to another. Therefore in this infinite set of universes, you are bound to get one universe (or more) where the ratios – the numbers – are just right. So rather than being a specially made universe, ours is just one of a huge number that happens to be configured so life can exist.

The problem with the multiverse theory is that it is based on very speculative physics, not the kind of physics we can do in the laboratory. Nor can we observe these universes. The question is: Why is our universe so special? The simplest answer – the one that satisfies science's requirement to choose the simplest of competing explanations – is that God made it. God made and designed the universe because he had a purpose for it – to bring about human beings who would be able to have a relationship with him.

*Scientists talk about the universe coming about through the Big Bang. They say that the universe expanded into being from a dense, hot state and is still expanding. Is the Big Bang, like multiverse, based on 'speculative physics'?*

No, there is a lot of evidence for the Big Bang. If a theory makes predictions and explains what is observed then that gives scientists a level of confidence in the theory. In the 1920s, the Belgian priest Georges-Henri Lemaître solved Einstein's equations of general relativity (his theory of gravity) applied to the universe as a whole and got, essentially, a Big Bang solution. Following Lemaître's work, Edwin Hubble observed that the galaxies were all moving away from each other – the universe was expanding. In other words, the observation backed the prediction.

Then in 1948 the American astrophysicist George Gamow argued that the early universe would have been very hot and dense and that matter and radiation would have been in equilibrium. But as the universe cooled, the matter and radiation would have decoupled, so today, 13.7 billion years after the Big Bang, there would be a cool, remnant radiation bathing the universe. In 1965 scientists detected this radiation and measured a temperature of three degrees above absolute zero. Again, observation followed prediction.

The Big Bang theory also correctly explains the abundance of helium. Some twenty-five per cent of the universe is helium. Astrophysicists were unable to explain the production of helium with models of chemical element formation in stars – the nuclear furnaces of the universe where most of the elements are made – but temperatures were also hot enough in the Big Bang for nuclear reactions to occur. Calculations show that just the right amount is made, so completing a satisfying account of how all the chemical elements were manufactured. Fred Hoyle was a major figure in all this work, especially stellar element production, but it is particularly ironic that he calculated how helium had been made in the Big Bang, when he was a proponent of the alternative steady-state theory and the person who coined the term 'Big Bang' as an insult.

The Big Bang theory is very well verified. We can reproduce conditions that pertained from one-thousandth of a second after the origin in our laboratories, and we have a good understanding of the physics from then on. For that reason the Astronomer Royal, Martin Rees, says that he is comfortable with what cosmology tells us from that point. However, the closer we get to the origin, the more speculative the science becomes. The most popular multiverse theories relate to much, much smaller fractions of a second than that, which is why I want to distinguish between what is well-established science and what is speculation, something which it is not very easy for laypeople to appreciate given the hype surrounding the more speculative theories, such as string theory.

*Science is full of theories – evolution, Big Bang, string, steady-state, chaos and inflation just to mention a few – which theories does the Bible support?*

The Bible gives us a *theology* of creation, not a *theory* of creation. It doesn't give us a scientific account like the Big Bang or the theory of evolution. A central theme is that God created out of nothing, so that everything that exists is totally dependent on God. The Bible also teaches the goodness of creation. It says that God created the universe freely, not out of compulsion. The universe is a result of God's motivation and choice. It could have been different from what it is, but God decided how it should be. God freely creating the universe is central to understanding life. If there is no Creator, everything falls to pieces – purpose, meaning, morality and ethics are nothing without God.

*Do you believe that science can explain the resurrection of Jesus?*

No, but as a scientist I would say the resurrection of Jesus is credible because there is powerful evidence for it. There is the empty tomb. There are the many appearances of the risen Christ. There is the testimony of the women he appeared to. In the first century – sadly

– women were not regarded as credible witnesses, so if you were going to invent this story, you wouldn't make women your witnesses. Then there is the rise of the Early Church – disillusioned, frightened disciples at the crucifixion, who a few weeks later were proclaiming the resurrection from the rooftops and were prepared to die for doing so.

*For some people, the resurrection and the likes of Jesus' subsequent appearance in a locked room are instances of God breaking his own scientific laws. How do you see them?*

It depends on how we understand the laws. Laws of science are just regularities that we observe. The reason that they pertain is because God is faithful in upholding those laws. God is not capricious. He doesn't break these laws all the time, otherwise that would result more in a Harry Potter world than one we can understand through science. But with the resurrection it looks as though the laws of God's new creation are breaking in.

*As a priest, you have met people on a much more down-to-earth level. You have counselled people in times of personal crisis. What insights, if any, from the scientific world are you able to offer people at a time of suffering? How do you answer the question: Why does a universe-creating God allow suffering?*

This is the most difficult question and the most serious objection to faith there is. There are no pat answers. When, as a priest, I meet people who are bereaved, it's not a time for theories. It is a time to weep with those who weep, as Paul says in Romans 12. It is a time to remember that the resurrection means that death is not the end for those who have faith in Christ. Science can give us an insight, though. Science describes a universe that shows us the necessity for birth and death and even extinction. The planet needs volcanoes and earthquakes – the causes of many deaths – in order to be fruitful.

The ultimate answer is in the cross of Christ. That is where God has taken the suffering of humanity on himself. The God of the Christian faith is not absent. He is not the giant experimenter in the sky who sets things going and has nothing to do with it afterwards. He is the God who is intimately involved with his creation, so much so that he has participated in it in the person of Christ. Jesus shared in human suffering; he experienced the loss of a friend, he experienced betrayal, the cruel death of the cross and – most of all – what it means to be God-forsaken. Jesus suffered all that, but God raised him to life, victorious over sin, death and pain – victory he is willing to share with those who follow him.

*After spending years studying the heavens, what do you think Heaven is like?*

I think it will involve a renewed physical creation, an embodied existence. There will be a measure of discontinuity and continuity. For instance, after the resurrection Jesus still had the marks of crucifixion on his body. His disciples still recognised him. The laws of physics will be different. The second law of thermodynamics won't hold: there will be no decay, the new body won't wear out. There will be a new Heaven. There will be a new Earth.

## FACT FILE

*The Reverend Dr Rodney Holder*
MA (Mathematics), Cambridge.
DPhil Astrophysics, Oxford.
BA (Theology), Oxford.
Fellow of the Royal Astronomical Society.
Fellow of the Institute of Mathematics and its Applications.

Operational research consultant with EDS (formerly Scicon) for Ministry of Defence 1980–1994.
Ordained as an Anglican deacon in 1997 and priest in 1998.
Curacy in Warwickshire 1997–2001.
Chaplain, English Church, Heidelberg, Germany 2002.
Priest-in-Charge, Parish of the Claydons, Oxford diocese 2002–2005.
Course Director of the Faraday Institute, St Edmund's College, Cambridge 2006–present.
1998 Templeton Foundation 'Exemplary Paper' Prize-winner for his paper on miracles.
Member of the Society of Ordained Scientists, the Science and Religion Forum and Christians in Science.
Author of *Nothing But Atoms and Molecules: Probing the Limits of Science* (Tunbridge Wells: Monarch, 1993) and *God, the Multiverse, and Everything: Modern Cosmology and the Argument from Design* (Aldershot: Ashgate, 2004).

# 16

# Discoveries

According to the science-fiction series *Star Trek*, space is the final frontier. In science fact there are, it seems, no boundaries. The thirst for knowledge is never satisfied. Whether to outer space or inner space, today's scientists boldly go where no one has gone before.

What fuels that journey? What makes a person spend thirty years studying two minutes of cell development? Well, there is money to be made. The first person to come up with a cure for the common cold, cancer or Aids would not only be immortalised but would also die a very rich person indeed.

But there is much more to scientific quest than fame and fortune. Among the things driving scientists are natural curiosity, a love of nature and the thrill of discovery.

For the scientist who is a Christian, there are the added attractions of recognising God in the universe he has created and, on discovering something new, the joy of being the first person after God to see it.

For some people the concept of a 'Christian scientist' is problematic. The scepticism tends to come from two opposite forces – non-Christians and fellow believers. In general, non-believing critics tend to think that a Christian scientist is not enough of a scientist. In a BBC *Newsnight* interview in October 2007, gene pioneer

Dr Craig Venter, who raced Dr Francis Collins to be the first to map the whole of the human genome, was asked if he was religious. In reply he said: 'I think it is impossible to be a true scientist and believe that there's a supernatural explanation for everything.'

(Some may question the premise that a religious person is one who has a 'supernatural explanation for everything' but for Venter faith and science don't mix.)

Perhaps the best-known modern-day critic of religion is Oxford professor, Richard Dawkins. In *The God Delusion*, he writes: 'The existence of God is a scientific hypothesis like any other.' Dawkins sees faith as a 'mind virus' and evolution, by definition, as atheistic. He cannot understand why Christian scientists say that they believe God created the world and that evolution is God's process for developing it.

In a live debate in Oxford with Professor Alister McGrath in March 2007, he said

> Most reputable theologians nowadays agree that life evolved by slow, gradual, incremental degrees. But they prefer to smuggle the Creator in as well, somewhat superfluously one might have thought. If natural selection, if evolution is God's way of designing life why would he choose the one way which makes it look as though he doesn't exist? Which makes his own role completely superfluous?

If non-Christians think that Christian scientists aren't scientific enough, then evolution is the measure by which some believers regard Christian scientists as not Christian enough.

A Christian scientist who says they believe that evolution is the way in which God sustains and develops his creation after the initial act – what scientists call a 'singularity' – is likely to be challenged as to the validity of their faith.

There is a school of thought that says that believing in God as Creator means having to take a literal view of the Genesis account of creation. Loosely, this is known as creationism.

In the past few years, creationism has been making headlines in the UK, particularly regarding whether it should be taught in schools. On 22 March 2002, it came up in Prime Minister's Questions when, *Hansard* records, Liberal Democrat MP Dr Jenny Tonge asked Tony Blair: 'Is the Prime Minister happy to allow the teaching of creationism alongside Darwin's theory of evolution in state schools?'

His reply began: 'I am very happy.'

On 19 October 2007, Radio 4's *Today* programme carried this story

> There's been a call for science teachers to respect the growing number of Christian and Muslim children whose families believe in creationism. That is a belief that the world was formed in its present state up to ten thousand years ago rather than more than four and a half billion years ago as most scientists believe. As our education correspondent James Westhead reports, the calls are contained in a new book.

James Westhead's report went as follows

> The book says schools in the UK are seeing a growing number of pupils – both Christian and Muslim – who believe in creationism, the biblical or koranic view that the world was created in six days. The author, a professor of science education, argues science teachers should take such beliefs seriously and respectfully, discussing

> them in class to help pupils reconcile the conflict with evolution. The view is controversial. Most teachers prefer to concentrate on science itself but Professor Michael Reiss [Professor of Science Education at the Institute of Education, University of London] argues that ignoring genuinely held belief could turn some children away from science altogether.

During the same month the National Curriculum guidelines on the teaching of creationism in schools were released. They included the following

> The theory of evolution lies at the heart of biology and should be taught at Key Stage 4 and in GCE advanced level biology. Creationism and Intelligent Design are not scientific theories and do not form part of the science National Curriculum or the GCSE and GCE A-level subject criteria. There may be situations in which it is appropriate for science teachers to respond to student comments or enquiries about the claims of creationism or Intelligent Design. This would be to establish why they are not considered as scientific theories... One way to do this would be to consider the mechanisms by which new scientific knowledge becomes established and why creationism and Intelligent Design do not meet these requirements.

Creationism has a number of strands. The most critical of evolution is Young Earth creationism. Proponents not only believe that God created the universe literally in the way described in Genesis 1 to 3 but also, by using the genealogy of Adam to Noah in Genesis 5, they calculate the universe to be no more than ten thousand years old. Therefore, they argue, evolution, which requires the universe to be billions of years old to allow for the gradual changes which have brought us to where we are today, is false. Ironically, this means that some self-confessed Bible-believing, evangelical Christians agree with self-confessed atheist, evolution evangelist Richard Dawkins. They both conclude that evolution is incompatible with the Christian faith.

It is not only the time aspect of evolution with which some creationists disagree. In their understanding, evolution teaches that humans descended from apes. This offends their belief in the truth expressed in Genesis 1:27: 'So God created man in his own image.'

There has long been speculation over a missing link between humans and apes. Victorian caricaturists often portrayed Darwin as a cartoon monkey.

Today's science tells us that humans and apes have some ninety-eight per cent of DNA in common. (Interestingly, although humans and mice share ninety per cent DNA in common, no one has seriously suggested that there's a missing link between mice and men.)

In the United States the conflict between creationism and Darwinianism infamously came to a head in July 1925 with the trial of biology teacher John Scopes, who was charged with teaching evolution in contravention of Tennessee's anti-evolution law. Section One of the Tennessee Evolution Statute 1925 stated: 'That it shall be unlawful for any teacher . . . to teach any theory that denies the story of the Divine creation of man as taught in the Bible, and to teach instead that man has descended from a lower order of animals.' Amid much news coverage, Scopes was found guilty but discharged on a technicality on appeal.

Most of the scientists in this book who address the subject say that evolution does not teach direct descent, but rather that humans and apes share a common ancestor.

The Christian scientists gathered in this book may or may not be representative of all Christian scientists.

(They are certainly not the only scientists who are Christian.) In approaching scientists for an interview, no cutting-edge sampling techniques were used to determine an accurate cross-section, nor did Darwin's theory of natural selection necessarily apply to the interviewee selection process. However, while the scientists were not questioned with the aim that they should be used as guinea pigs or lab rats, some common conclusions have emerged.

## Creation

The scientists in this book share a common view of why there is something rather than nothing. Unanimously, they believe that God created the universe. It is important to them that he did and that life is not the result of random chance. It is important to them that humankind is made in God's image, as Genesis 1:26, 27 states. In a nutshell, they combine insights from their faith and their science by holding that God created the universe through Big Bang and that he sustains it through evolution. They also believe that God is continually creating. Therefore, they regard themselves, in some sense, as creationists.

This is not to say that they believe in the ideas of creationism mentioned above. Nor should it come as a surprise that they are happy with the 'creationist' label. Affirmations of faith recited by millions of believers around the world every day begin with statements that God is the Creator. The Apostles' Creed begins: 'I believe in God, the Father Almighty, the Creator of Heaven and Earth.' The Nicene Creed, formulated in AD 325, commences: 'We believe in one God, the Father Almighty, maker of all things visible and invisible.'

Old and New Testaments contain many references to God creating the universe and everything in it, as this quick sample indicates

> You alone are the LORD. You made the heavens, even the highest heavens, and all their starry host, the earth and all that is on it, the seas and all that is in them. You give life to everything, and the multitudes of heaven worship you (Neh. 9:6).

> In the beginning you laid the foundations of the earth, and the heavens are the work of your hands (Ps. 102:25).

> He [Jesus] is the image of the invisible God, the firstborn over all creation. For by him all things were created: things in heaven and on earth, visible and invisible … all things were created by him and for him (Col. 1:15,16).

> By faith we understand that the universe was formed at God's command, so that what is seen was not made out of what was visible (Heb. 11:3).

The Bible, say the scientists – with particular reference to the opening chapters of Genesis – is not a scientific textbook. As Rodney Holder puts it: 'The Bible gives us a *theology* of creation, not a *theory* of creation.' To read it as a textbook, they say, is to do Holy Scripture a disservice. To them, biblical accounts of creation are more about the 'Who?' and the 'Why?', than the 'How?' and the 'When?'

## Limitations of science

One of the challenges all Christians face is naturalism. According to naturalism, the world can be explained without reference to the spiritual or supernatural, because the spiritual and supernatural do not exist. Under such a philosophy, it is science that provides all the answers.

It is easy to see how this could happen. Over the years, we have treated white-coated scientists with awe. Science has what John Lennox calls 'immense cultural authority'. When we are so overwhelmed by technicalities we lesser mortals talk of being 'blinded by science'. We set great store by the word of physicians. If we miss work for a more than a week because of illness, our employer requires verification from our GP. It is doctors who have to sign to allow a termination of pregnancy and it is doctors who verify death on a death certificate.

On the larger scale, medical science can be said to be working miracles. Things considered impossible fifty years ago – body-part transplants and fertility treatments, for example – are now everyday procedures. It is unlikely to be long before gene therapy will become more widespread. Thanks to science, death is being cheated and life created.

We are dazzled by technology. Our homes and our lives are filling up with iTech. We have Wii, wi-fi, MP3s, CDs, DVDs, GPS, ABS and HDTV. Computer scientists and technicians talk about 'solutions' – printing solutions, network solutions, data storage solutions . . . The subliminal message seems to be: Got a problem? Science will solve it.

But, say our scientists, there are some solutions that science can't provide. There are questions that science will never answer because science is limited to the natural world. Questions such as 'Why am I here?'or 'What is the meaning of life?' and 'Why is there something rather than nothing?' do not have scientific solutions. They need another approach – the perspective of faith.

We may live in a digital age but we still need the hand of God.

## Faith

Before we consider some of those questions, let's look at the perspective of faith and what our scientists mean by faith. Unanimously, they reject Richard Dawkins's working definition of faith as 'blind trust' or as a 'running away from evidence'.

Christianity is a faith rooted in fact. Christianity is a relationship with a real person. Christianity involves an eternal God stepping into human space and time in the person of Jesus. As Paul puts it: 'But when the fullness of time had come, God sent his Son, born of a woman, born under the law, in order to redeem those who were under the law, so that we might receive adoption as children' (Gal. 4:4,5).[4]

Each of the Gospel writers refers to contemporary events, places and people – facts. Again, a non-exhaustive survey may prove the hypothesis.

In his nativity narrative, Matthew says that Jesus was born 'in Bethlehem in Judea, during the time of King Herod' (Mt. 2:1). Herod reigned from 40 to 4 BC. Matthew also chronicles the journey of the Magi who saw a star in the east. In his interview, Professor Colin Humphreys shares his research on the star of Bethlehem. Chinese astronomical records, he says, show that the star – a long-tailed comet – was real and appeared in the region in 5 BC for seventy days.

In the opening verses of Mark's Gospel, Jesus 'came from Nazareth in Galilee and was baptised by John in the Jordan' (Mk. 1:9).

Luke – a physician (does this make him the first Christian scientist?) – anchors his Gospel in a number of contemporary events. According to Luke, John the Baptist begins his ministry 'in the fifteenth year of the reign of Tiberius Caesar – when Pontius Pilate was governor of Judea, Herod tetrarch of Galilee, his brother Philip tetrarch of Iturea and Traconitis, and Lysanias tetrarch of Abilene' (Luke 3:1). Tiberius's successor, Augustus, died in AD 14, so Tiberius's fifteenth year would be 29 AD. In Luke 13:4, Jesus mentions a piece of local news, when he refers to 'those eighteen who died when the tower of Siloam fell on them'.

While John's Gospel appears more theological than the synoptic Gospels, the evangelist does put Jesus into real time, notably by marking his ministry by three annual Passover feasts

> When it was almost time for the Jewish Passover, Jesus went up to Jerusalem (Jn. 2:13).

> Then Jesus went up on a mountainside and sat down with his disciples. The Jewish Passover Feast was near (Jn. 6:3,4).

> Six days before the Passover, Jesus arrived at Bethany, where Lazarus lived, whom Jesus had raised from the dead (Jn. 12:1).

There is one independent, historically verified person who is mentioned by all four Gospel writers – the Roman procurator of Judaea, who authorised the crucifixion of Jesus – Pontius Pilate. Pilate, according to New Testament scholar, the Reverend Professor Richard Burridge, was 'a Roman knight (the second rank, below senators) and governor of Judea AD 26–36'.[5]

The Roman historian Tacitus mentions Pilate and Christ in the same sentence when explaining the origins of Christians being persecuted by Nero in Rome in the 60s AD

> Nero fastened the guilt and inflicted the most exquisite tortures on a class hated for their abominations, called Christians by the populace. Christus, from whom the name had its origin, suffered the extreme penalty during the reign of Tiberius at the hands of one of our procurators, Pontius Pilatus (*Annals* 15.44).

References to Jesus by other external historians such as the Jewish Josephus (circa AD 37–100) and Romans Pliny and Suetonius – plus hostile mentions in rabbinic writings – support the fact that there was an historical figure called Jesus who lived where and when the Gospel writers describe.

For Christians, however, Jesus of Nazareth is more than a historical, factual figure. Christians believe that through his sinless life, death and resurrection, Jesus is the way by which God forgives people their sin and gives them eternal life.

When discussing whether or not there is an Intelligent Designer behind the universe, scientists often talk of irreducible complexity. By this they mean that something is made up only of essential components and that if one of these were taken away, the whole thing would cease to work. Does Christianity have an irreducible complexity? Perhaps the apostle Paul comes close when he writes: 'By this gospel you are saved, if you hold firmly to the word I preached to you . . . that Christ died for our sins according to the Scriptures, that he was buried, that he was raised on the third day according to the Scriptures, and that he appeared (1 Cor. 15:2–4).

Here are four things that a person needs to believe about Jesus if they are to be a Christian: Jesus died for our sins, Jesus was buried, Jesus was raised and after his resurrection Jesus appeared. But Christianity is not only to do with believing things *about* Jesus. It also requires believing *in* Jesus. Put simply, a Christian is someone who trusts that the death of Jesus – in their place, for their sin – is the means by which God will give them a place in Heaven.

A personal faith in Christ is not, as Dawkins likes to put it 'blind trust'; it is a blend of fact and faith – head and heart. For Christians, Jesus is also the cosmic Christ. They believe that Jesus existed and exists eternally – in other words, outside of time.

Like the first bars of an overture, John's Gospel opens with a series of attention-grabbing statements

> In the beginning was the Word, and the Word was with God, and the Word was God. He was with God in the beginning. Through him all things were made; without him nothing was made that has been made. In him was life, and that life was the light of men (Jn. 1:1–4).

The theme of an eternal Christ continues in the writings of Paul, who writes: 'He is before all things, and in him all things hold together' (Col. 1:17). Peter adds: 'He was chosen before the creation of the world' (1 Pet. 1:20).

Christians believe that in the incarnation, God in the person of his Son, Jesus, stepped from eternity into time. Christians believe that, at his ascension, Jesus stepped from time into eternity and that, at a time of the Father's choosing, Jesus will again step into time and return to the world. So then, for our scientists, faith in Jesus Christ is not a blind trust in some mythical figure such as Hercules, Santa Claus or the Tooth Fairy. It is faith in a person for whom there is evidence. It is a faith based on facts.

It is also a personal faith. The journeys that these men took to faith in Christ are all individual. Alister McGrath and Francis Collins came to faith from atheism. Others, such as Sir Ghillean Prance, grew up in churchgoing families but were only nominal Christians. Some, Russell Stannard included, describe their coming to faith as a process, while Michael Poole can recount his conversion in time, place and date format. But, whatever the circumstances of the new birth, each scientist testifies to being alive in Christ. In this, again, our scientists are typical of the wider Christian family. God treats us as individuals. He also deals with us as people and not on account of our qualifications or professional standing – or lack of them.

There is another element of faith which our scientists highlight: all scientists need faith – although not all will label it as such – in the scientific process. Science is not sterile. John Polkinghorne puts it like this: 'Scientists need to commit themselves to the belief that the physical world is intelligible and that there are helpful ways of thinking about it.' Scientists, says Denis Alexander

> often don't realise how much implicit faith they have. For example, a scientist has to believe that the universe is rational and reproducible – that it has properties that can be investigated in a rational way: that if they do an experiment this week, it will produce the same results next week. There are all kinds of implicit background understandings of 'faith' that all scientists have, whether they like it or not.

Scientists are not hermetically sealed. They take into their labs what Hungarian physicist Michael Polanyi calls a 'fiduciary framework' – a framework of faith. He writes

> We must recognise belief once more as the source of all knowledge. Tacit assent and intellectual passions, the sharing of an idiom and of a cultural heritage, affiliation to a like-minded community: such are the impulses which shape our vision of the nature of things on which we rely for our mastery of things. No intelligence, however critical or original, can operate outside such a fiduciary framework.[6]

The idea, then, that religious faith is all emotionalism, whereas science is one hundred per cent pure rationalism, is a convenient but inaccurate caricature.

## Miracles

Our scientists agree that some questions cannot be answered by science. One such area of life is miracles. All the scientists agree that miracles can and do happen. But what is a miracle?

Michael Poole says it is 'a surprise event which is a sign from God that attracts our attention and causes us to wonder about him'. For Francis Collins, a miracle is God's 'invasion of nature', while Chris Southgate sees a miracle as 'a unique event which particularly displays the character of God in relation to the world'. Bob White defines a miracle as 'something outside normal scientific happening'.

So where does miracle fit with science? Alister McGrath has an answer

> It is commonly said that science disproves miracles. It doesn't. Science says that there are certain things that by their nature are very improbable. So improbable, in fact, they shouldn't happen at all. Science can't say

> things can't happen, merely that they are very improbable.
>
> For the Christian, miracles seem to have this tendency to happen around Jesus. For the Christian, therefore, modern science makes these events all the more remarkable and forces us to ask the question: 'What is so significant about this man who does all these things?'

Colin Humphreys says that when he reads the Bible, he tends to look for natural explanations: 'That doesn't mean I don't believe in miracles but I think God often works in and through nature to achieve his purpose.' On these occasions, he says, 'the miracle is often in the timing of an event'. Humphreys also gives a different slant to the scientific approach to miracles. Speaking of Jesus walking on water, he says: 'Rather than a natural phenomenon [such as gravity] being suspended, why not say God provided an additional force which upheld Jesus? That way God doesn't break his own rules to perform a miracle.'

No matter how closely scientists try to apply their science to miracles, when it comes to the virgin birth and the resurrection, they readily admit that science has no answers or explanations. Molecular biologist John Bryant summarises opinion on the virgin birth: 'I can't explain the virgin birth, either as a scientist or as a Christian. It can't have been a divine cloning experience or Jesus would have been female. It is a matter of faith that Jesus was born divine and human.'

Science, says Alister McGrath, 'does not disprove the resurrection of Jesus. It says that the resurrection cannot be explained by a natural process. The question is not so much "How did the resurrection happen?" but "What does it mean?"'

For John Lennox, the meaning of the resurrection is clear. 'The resurrection is a new act of creation. It shows that there is a destiny beyond death and that those who choose to do so will share in that life.'

Michael Poole states the situation succinctly: 'The resurrection is not a scientific question.'

## Suffering

There is one question that puzzles even the keenest of scientific minds: Why does an all-powerful, all-loving God allow people to suffer? Our scientists agree that this is a tough one.

'This is the most difficult of all questions,' says John Polkinghorne. 'It is one which holds people back from faith and one which causes some believers to lose faith.'

A partial answer for Polkinghorne lies in the evolutionary process – or what he calls 'a world allowed by the Creator to make itself'. He explains

> What produces new forms of life is genetic mutation. But if you have genetic mutation, some cells can mutate and become malignant. You can't have one without the other. So, for example, the existence of cancer in the world is the shadow side of the fruitful history of the world ... The more science understands the world, the more it seems a sort of package deal, that there is inevitably a good side and bad side to many things.

Cancer specialist Denis Alexander also interprets suffering as part of a package deal: 'With carbon-based life you have to have transience – you have to have life and death. It is built into the system – life without death makes no sense – that's the package deal . . . The downside of genetic variation is disease, because genetic variation sometimes causes disease – including cancer.'

If that sounds a little too cold and clinical, he also points out that 'God has a new Heaven and a new Earth in mind for us. Only against the backdrop of a new Heaven and new Earth can suffering start to make sense.'

Climate change expert Sir John Houghton points out that natural disasters – floods and droughts – are the greatest cause of deaths and commercial loss. Some suffering, he says, is caused not by God but by humankind. 'Human greed,' he says, 'is at the heart of global warming.' He continues

> God has placed us in a world which we can get to know and understand. We can learn to predict and deal with floods and droughts. We can learn to alleviate the consequences of floods, drought and earthquakes. The fact that we sometimes don't is not God's fault. It is our fault. The fact that we don't radically help those who constantly live with floods or drought is not God's fault. It is our fault.
>
> That doesn't answer personal tragedy, when we call out: How on earth could God allow that to happen? And I ask that question, as much as anyone else. But I have to leave the answer to God.
>
> What I do know is that God sent his Son into an imperfect, sinful world to suffer and die in order to redeem the world. The ultimate answer to sin, suffering and pain is the cross of Jesus. I don't understand the cross but I accept its power.

It is to the realm of faith that our scientists turn in coming to grips with the question of suffering. 'Suffering, pain, disease, illness are all consequences of man's initial disobedience to God (the fall) – part of man's nature,' says physician Norman Nevin.

'One of the essential Christian insights into suffering,' according to John Polkinghorne, 'is that God is not just a compassionate spectator. But that in the cross of Jesus Christ, God has participated from the inside in the suffering of the world. The Christian God is the crucified God as well as the redemptive God.'

But ultimately, the question of suffering – like many a scientific and mathematical problem – seems insoluble. Alister McGrath sums it up

> I don't think we're ever going to explain suffering completely. Christianity doesn't offer us a neat theory of suffering. It speaks to us of a God who is present in suffering … How did God redeem the world? Through suffering – the suffering of Christ, his Son. It was a suffering that one day will bring about the complete elimination of suffering from the world. There are no easy explanations but there is this reassurance of a God who has been through suffering and will be with us as we suffer.

And it is a relationship with a God who suffers with his creations that is at the heart of the Christian faith.

## Proving God

Science – and maths – are known for their proofs. In law, forensic evidence is produced to prove a defendant's innocence or guilt. In pub-table discussions, 'but science proves it' is used to trump all other arguments.

Maybe it goes back to the 'immense cultural authority' John Lennox says we give science, but it seems we accept scientific proof as the final word. However, scientific proof may not be as nailed-down or watertight as we have been led to believe. Particle physicist and broadcaster, Dr Simon Singh, describing the nature of scientific proof in his book *Fermat's Last Theorem*, explains

> In science a hypothesis is put forward to explain a physical phenomenon. If observations of the phenomenon compare well with the hypothesis, this becomes evidence in favour of it. Furthermore, the hypothesis should not merely describe a known phenomenon, but predict the results of other phenomena. Experiments may be performed to test the predictive power of the hypothesis, and if it continues to be successful then this is even more evidence to back the hypothesis. Eventually the amount of evidence may be overwhelming and the hypothesis becomes accepted as a scientific theory.
>
> Scientific theory ... is merely considered highly likely based on evidence available. So-called scientific proof relies on observation and perception, both of which are fallible and provide only approximations to the truth ... Even the most widely accepted scientific 'proofs' always have a small element of doubt in them.[7]

When it comes to scientific proof and faith, our scientists have no doubts at all: science can neither prove nor disprove the existence of God. Professor Richard Dawkins disagrees. In his public debate with Alister McGrath at Oxford in March 2007 he said

> I believe that the question of the existence of God or gods, supernatural beings, is a scientific question, whereas other scientists will say it's nothing to do with science, science and religion occupy two quite separate majesteria and don't overlap. I think they do overlap, I think they both attempt to answer the same kinds of questions. The difference is that religion gets the answers wrong.

John Polkinghorne thinks otherwise: 'I don't think you can prove the existence of God. In fact, I don't think you can even prove the existence of quantum things like electrons. It seems to me, you can have good reasons for believing in them but you can't prove them . . . We can't prove God but we can experience him.'

In Michael Poole's opinion, 'The existence of God is a faith question – and one for which Christian faith gives an answer – and not a scientific question.'

Wanting some sort of proof of God is, of course, not new. In the Gospels, Jesus' audience often ask him for a sign – some token of divine authority – as this sample shows

> The Pharisees and Sadducees came to Jesus and tested him by asking him to show them a sign from heaven (Mt. 16:1).

> Then the Jews demanded of him, 'What miraculous sign can you show us to prove your authority to do all this?' (Jn. 2:18).

> Jesus answered, 'The work of God is this: to believe in the one he has sent.' So they asked him, 'What miraculous sign then will you give that we may see it and believe you? What will you do?' (Jn. 6:29,30).

There is one subscriber to the seeing-is-believing school of thought who has become so well known that his name has fallen into common parlance – Doubting Thomas

> The other disciples told him, 'We have seen the Lord!' But he [Thomas] said to them, 'Unless I see the nail marks in his hands and put my finger where the nails

were, and put my hand into his side, I will not believe it' (Jn. 20:25).

Preachers often give Thomas a hard time – 'How could he not believe? Why couldn't he have taken it on trust? He should have had more faith!' But there is so much that is positive in this encounter. Thomas is an ordinary bloke who wants proof of Jesus' resurrection. Jesus understands Thomas. Jesus does not reject Thomas for his lack of faith. Jesus gives Thomas all the proof he needs, in a way he can understand it.

Could this be a template – the divine method – for proving Jesus for ourselves? Although it is beyond science to prove God, our scientists point to a number of scientific phenomena as signs – evidences – that there is a God.

### *The Goldilocks Effect*

The first is that life on Earth is possible only because the universe is finely tuned. Scientists call this the anthropic cosmological principle. Non-scientists may prefer the 'Goldilocks Effect'. Out of the one hundred billion galaxies, each comprising one hundred billion stars that make up the universe, conditions are – like Goldilocks's porridge – 'just right' for life on Earth. Geophysicist Bob White makes the generally agreed point

> The universe is extremely finely balanced. The anthropic principle, which is a subject of scientific study, is concerned with some of these incredibly precisely tuned parameters that make life possible. Someone once suggested that the tuning of some of the physical constants is as accurate as hitting a one centimetre square target with an arrow fired from the other side of the universe.

> There has to be the right force of gravity, the right forces of attraction and repulsion in the nucleus, the right electrical forces and so on, otherwise matter as we know it, and ultimately life itself could not exist. One way of looking at that is to say that we're here just by chance. But as a Christian I see it as evidence that a gracious and loving God created a place just right for us to inhabit.

Referring to the cosmic archery metaphor, Rodney Holder reasons: 'The question is: Was that a lucky shot or was it a brilliant marksman? Likewise, is the fine balance of the universe a lucky happenstance or is there a brilliant designer behind it? I believe that there's a brilliant Designer behind it.'

### *Dark Energy*

It is not only Christian scientists who recognise that the universe is finely tuned. Cosmologist Professor Paul Davies, who describes himself as 'not a person of religion', is director of the Beyond Research Centre at Arizona State University. In his book *The Goldilocks Enigma,* he highlights a number of areas of cosmic fine-tuning through the four forces of nature: gravity, electromagnetism, weak and strong nuclear force. But, with the universe expanding, it is dark energy that particularly excites Davies. He writes, 'Dark energy is a cosmic repulsion, or anti-gravity force, which drives the galaxies apart at an accelerating rate.'[8]

Dark energy, it seems to me, works something like this: Imagine the universe as a car. Dark energy is what happens when you press the accelerator pedal – you move further away from your starting point. The car also has a brake. This is normal attractive gravity. Applying the brake, of course, either slows the car or brings it to a complete halt. The driver of our car, though, drives while applying the accelerator and brake at the same time.

The question is, if the car is going forward (the universe is expanding) how much more accelerator is being applied than brake? The answer, says Davies, is one part in $10^{120}$ (10 times 10, 120 times).

'Our existence depends on the dark energy not being large,' he writes

> A factor of ten would suffice to preclude life: if space contained ten times as much dark energy as it actually does, the universe would fly apart too fast for galaxies to form. A factor of ten may seem a wide margin, but one power of ten on a scale of 120 is a pretty close call. The cliché that 'life is balanced on a knife-edge' is a staggering understatement in this case: no knife in the universe could have an edge that fine.
>
> Logically it is possible that the laws of physics conspire to create an almost but not quite perfect cancellation. But then it would be an extraordinary coincidence *that* that level of cancellation … just happened by chance to be what is needed to bring about a universe fit for life. How much chance can we buy in scientific explanation? One measure of what is involved can be given in terms of coin flipping: odds of $10^{120}$ to one is like getting heads no fewer than four hundred times in a row. If the existence of life in the universe is completely independent of the big fix mechanism – if it's just a coincidence – then those are the odds against us being here. That level of flukiness seems too much to swallow.

While Paul Davies does not make a connection from 'level of flukiness' to God, our scientists argue that rather than being the result of some random happenstance, the universe is designed and – therefore – has a Designer.

### *The moral law*

The next piece of evidence our scientists point to is the moral law. Francis Collins explains

> The moral law cannot be readily explained in terms of evolution because it sometimes calls us to do things that are really quite the opposite of what evolution would ask – like jump into a river to save a drowning stranger. The moral law has been fully set in the heart of only one species – humankind. There's no bigger signpost towards the existence of a personal God.

Other pieces of evidence for God offered by the scientists include the existence of a worldwide faith community, the existence of art and music, a night sky, the beauty of mathematics, the wonder of DNA and the fact that humans are made from the dust of dead stars. But the ultimate evidence, they agree, is not a scientific proposition. It is a person. It is Jesus.

Where many people have problems with Jesus is not that he lived and died (external sources as well as the New Testament provide evidence for that); it is that God raised him to life. Those looking for evidence of the resurrection might like to consider the following:

### *Written evidence*

All four Gospels describe the death, burial and resurrection of Jesus. The earliest, Mark, appeared around thirty years after Jesus' death. Existing complete copies of the Gospels date from around AD 340. By comparison, earliest copies of books such as Julius Caesar's *The Gallic Wars* date from many more hundreds of years away from the events they describe. If we can readily believe the accounts of Caesar's campaign in France, why dismiss the Gospels as less than historically accurate?

### *Eyewitness accounts*

All four Gospels record accounts of people meeting Jesus after his resurrection. The witnesses listed are: Mary Magdalene (Mt. 28:9, Mk. 16:1, Jn. 20:16); Mary (Mt. 28:9, Mk. 16:1); eleven disciples (Mt. 28:16–17, Mk. 16:14, Lk. 24:36); Salome (Mk. 16:1); Cleopas plus one (Lk. 24:18); Simon (Lk. 24:34); ten disciples (Jn. 20:19); Thomas (Jn. 20:26); Simon Peter, Thomas, Nathanael from Cana, James and John plus two (Jn. 21:2); Joseph Barsabbas and Matthias (Acts 1: 22–26); Peter (1 Corinthians 15:5); the Twelve (1 Cor. 15:5); more than five hundred unnamed disciples (1 Cor. 15:6); James (1 Cor. 15:7); all the apostles (1 Cor. 15:7) and Paul (1 Cor. 15:8).

Even allowing for multiple mentions, the New Testament says that at least five hundred and twenty people saw Jesus after he was raised. Under Jewish law at the time of Jesus, a criminal conviction could be decided on the testimony of just two or three witnesses (Deut. 19:15). Yet there are thirteen named witnesses to the resurrection of Jesus. What would today's police give for thirteen identifiable witnesses to an incident?

These witnesses gave oral evidence. After the resurrection, the disciples told people that the once-dead Jesus was now alive. Why didn't the authorities put a stop to it by producing his body?

Three of the first people to witness the empty tomb on Easter morning were women – Mary, Mary Magdalene and Salome (Mk. 16:1). Under Jewish law at the time, a woman's testimony was not regarded as binding. If the resurrection was a story concocted by the disciples – a fabrication – would they have cast women among the primary witnesses, knowing that their word would be seen as less reliable than a man's?

Some close disciples – for example, Stephen (Acts 6 and 7) and James (Acts 12:2) – faced death because they refused to stop preaching that Jesus was alive. They refused the chance to recant.

In AD 64, the Roman emperor Nero started the first mass persecution of Christians. There have been many persecutions and pogroms since. Christians, for example, were persecuted and slaughtered during the days of the Ottoman Empire and the Ming and Edo dynasties. In the twentieth century, many Christians living under Communism, Fascism or National Socialism were killed because of their faith – a faith whose indispensable condition is the resurrection of Jesus. Faced with the stark reality of one's death before sunrise, is a known lie really worth dying for?

### *Forensic evidence*

After his death Jesus was put in an unused tomb (Jn. 19:41). It was sealed with a stone (Mt. 27:60), which was itself also sealed (Mt. 27:66). There was a round-the-clock military guard (Mt. 27:66).

Jesus was the first and only occupant of the tomb. When, therefore, on Easter morning the stone is rolled away, the guard has vanished and the tomb is empty, it cannot be a clerical error. There was not – there never had been – another body in the tomb. It was not a case of mistaken identity. This is a case with no corpse.

Before his burial, Jesus was hastily dressed in linen (all four Gospel writers mention the fabric) grave clothes. On Easter morning, as Luke and John record, the clothes were left in the tomb. John's Gospel records that the cloth that had been wrapped around Jesus' head was folded (20:7). Folded!

If his disciples had crept into the garden of Gethsemane undetected in the middle of the night, overcome the armed Roman guards, broken the seal, rolled the stone away and carried the mutilated corpse of Jesus away, why would they spend time – and risk detection and certain death – by removing his clothes and then folding them? Who tidies a tomb? And if not the disciples, then who moved the stone?

Who would be the suspects? The holy trinity of fictional and factual whodunnits is means, motive and opportunity. An alternative explanation to the Gospel account of resurrection has to satisfy the question of means, motive and opportunity. If not God, then who?

Three grief-sticken women at the tomb on Easter morning? The traumatised disciples, whose own initial reaction to the women's report of the resurrection was that it was an 'idle tale' (Lk. 24:11)?[9] The headstrong disciple Simon Peter who was guilt-ridden after denying Jesus three times? The guards watching the tomb? The Jewish authorities? Surely if they'd taken the body they would have produced it as soon as his disciples started preaching that Jesus was raised from the dead. Who of them had the means, motive and opportunity to roll back the stone, remove the body and keep it hidden?

In his book *Jesus the Jew* Geza Vermes, Emeritus Professor of Jewish Studies, Wolfson College, Oxford, writes

> In the end, when every argument has been considered and weighed, the only conclusion acceptable to the historian must be that the opinions of the orthodox, the liberal sympathizer and the critical agnostic alike – and even perhaps of the disciples themselves – are simply

> interpretations of the one disconcerting fact: namely that the women who set out to pay their last respects to Jesus found to their consternation, not a body, but an empty tomb.[10]

In seeking explanations for their observations, scientists often apply a principle known as Occam's razor. It says that when confronted with a set of competing explanations, choose the simplest.

Many followers of Jesus have done just that. They've looked at the evidence and concluded that God raised Jesus from the dead; they've asked Jesus to prove himself and experienced the life-changing power of his resurrection for themselves.

The final word goes to Russell Stannard

> Nobody ever gets argued into a loving relationship with God. Science neither proves nor disproves his existence. The strongest evidence for God comes from your own experience, what you get out of your relationship with him. That is something a person has to try for themselves. Unless you have honestly tried to pray, to enter into that relationship and sense the presence of God, then arguing is a waste of time.

And that, surely, applies however much we know about science or Christian scientists.

# Endnotes


[1] Richard Dawkins, *The God Delusion* (London: Bantam Press, 2006).

[2] St. Augustine, *The Literal Meaning of Genisis*, Vol. 1, translated by J.H. Taylor (New York: Paulist Press, 1982).

[3] John Lennox, *God's Undertaker: Has Science Buried God?* (Oxford: Lion Hudson, 2007).

[4] *New Revised Standard Version.*

[5] Richard Burridge, *John: The People's Bible Commentary* (Oxford: BRF, 2007).

[6] Michael Polanyi, *Personal Knowledge* (London: Routledge & Kegan Paul, 1958).

[7] Dr Simon Singh, *Fermat's Last Theorem* (London: Fourth Estate, 1998).

[8] Paul Davies, *The Goldilocks Enigma* (London: Penguin, 2007).

[9] *Revised Standard Version.*

[10] Geza Vermes, *Jesus the Jew* (London: Collins, 1973).